Travelling by sound and smell

My world of aircraft, airlines, airports and attractions

Travelling by sound and smell:
My world of Aircraft, Airlines, Airports and Attractions.

Copyright 2023 John CT Miller

Published by Piggle Design

The moral rights of the author have been affirmed
and all rights reserved.

You may not copy, distribute, transmit, reproduce or otherwise make available this publication or any part of it in any form or by any means including without limitation electronic, digital, optical, mechanical, photocopying, printing and recording or otherwise without the prior written permission of the publisher. Any person, who does any unauthorised action in relation to this publication, may be liable to criminal prosecution and civil claims for damages.

Condition of Sale

This book is sold subject to the condition that it shall not, by way of trade or otherwise, be lent, re-sold, hired out or otherwise circulated in any form of binding or cover other than that in which it is published and without a similar condition including this condition being imposed on the subsequent purchaser.

Disclaimer

This book is a work of fiction. Names, places, characters, and events are the product of the author's imagination. Any resemblance to actual persons, living or dead, events, or locales is purely coincidental.

John CT Miller is a former South African journalist, who previously covered and wrote about entertainment, courts, crime, and consumer affairs, before settling in the UK and turning his hand to writing books.

Prologue

Rudyard Kipling said "the first condition about understanding a foreign country is to smell it."

Well, I have to agree, but in my case, it was not just about smell, but sound as well as gastronomy.

Sitting back in my leather seat travelling at twice the speed of sound at 60000 feet above the Atlantic ocean along with 100 other passengers on the supersonic Concorde I could hardly believe this was happening to me. Aircraft and that little kid me had both definitely come a long way.

As a child, even in my wildest dreams I never imagined this could happen.

For many years, breaking the sound barrier or Mach1 was just a phrase and reality reserved for fighter jet pilots.

In my youth, I also had no idea smell, sound and strange tasting foods would be part of my travelling world.

But before you board and get ready for take-off, I must remind you that none of this book may be copied in any form without the permission of the publisher Catherine@piggle.design.co.uk or you can contact me direct at johnctmiller@gmail.com

You can also send any comments to Catherine or me.

I must thank Catherine Murray once again for the front cover design and the formatting of the e-book and paperback editions as well as a fellow aviation journalist Linden Birns, who guided me through any turbulence while writing this book. We did several trips on various aircraft together.

Chapter 1

When I was very young, I dreamed of becoming a jet fighter pilot. in those early days I never thought when building and collecting model aircraft my passion for planes would lead me years later to first flying on board the almost 90 year old Ju 52 through to Concorde at a speed of Mach 2.4, as well as many other aircraft types in-between.

Back in the 50s and 60s passengers if they were lucky were offered a newspaper to read, while today some airlines offer more than 5000 inflight entertainment channels.

Before the days of the internet, cell phones, computer games and social media many young children grew up playing with and probably trying unsuccessfully like me to build model aircraft.

Those small balsa wood models needed patients to assemble them, and as a 7 or 8 year old, patience was never much of a virtue.

I guess like me, many gave up half way, and left it to their dad or an uncle to complete the half assembled plane.

I also bet many little boys like me probably spent more time outside playing cowboys, cops and robbers, firemen, or climbing and swinging from trees.

In those early days, the closest I got to flying was certainly not by choice, but by accident. This happened when I fell out of a tree, leaving me with cuts and bruises.

The short lived flights were not just falling from trees, but flying out of our home-made go-cart.

John CT Miller

These unintended flights happened when my brothers and I took it in turns to race down the 45 foot (15 metre) ant hill or termite mound in our garden.

Our own termite mound had several trees growing around and on top of it.

The termite mound is still visible on Google Earth.

We often played and practised swinging from tree to tree or seeing how far up a tree we could clamber without falling out of the tree.

Each branch or limb had first to be tested for strength, while at the same time looking out for snakes.

A misguided foot or hand could find one of us flying uncontrollably to the ground.

The go-cart or contraption was extremely basic. 4 fixed wheels and a screwed on piece of wood to sit on.

There was no stearin wheel, relying more on luck to make sure you didn't fall out, or fly off it.

From the top of the ant hill our crude racing track was extremely uneven and included a couple of very uneven man-made steps.

We never knew which bumps or steps the go-cart would take or hit as each of us made our rapid descent.

The game plan with one of us sitting in it was to see how far the go-cart would go from the top of the termite mound until it came to a stop somewhere in our garden.

Whoever stayed on board and went the furthest was the winner.

The challenge was not only to still be on board when it finally stopped, but avoid flying off it as it made its unpredictable way down the ant hill.

In those days, there was no such thing as protective clothing.

Today, I am sure such games would be prohibited by health and safety authorities as well as over-protective parents.

Travelling by sound and smell

If there weren't shrieks of laughter or tears coming from the garden, parents often wondered about the sounds of silence and where we were, or what mischief we were getting up to.

The old folks suspicions were most times proved right. The silence probably meant we were doing something which we weren't supposed to do.

I often ponder if it is happenstance that defines later actions and experiences, or is it deliberate planning or even divine providence?

I guess I was not the only little boy, who dreamed of becoming a pilot. Not just a civilian passenger pilot, but a glamorous jet fighter pilot: able to play games in the skies.

How many of us in those days were excited when mom and dad promised to take us over the weekend to the local airport to watch planes landing and taking off? This trip to the airport was known as a family outing or much more a treat for young would-be aviation enthusiasts.

I am sure those in families who had no interest in watching planes at the airport, the anticipation of that special milkshake, a brown cow, and a cream horn and if you were old enough even a rum baba would make up for the hour or two of boredom.

In those days there were no such things as cupcakes or fancy coffee or even sugar-free or diet sodas.

These were not only the days before modern technology, but also before strict security measures at airports.

Back then, families were allowed to sit outside on the airport roof or balcony and from a few metres away watch the planes.

How many parents realised weekend afternoons were probably not the best time to see many planes? Certainly in Africa in the early 60s, there were not that many flights especially over weekends.

For any young plane enthusiast, the milk shake and slice of cake was quickly forgotten for a few minutes as they watched the man made birds of the sky take off and land.

I wonder how many little girls and boys hoped one day they might be a passenger let alone a pilot?

Who would have ever imagined some 50 years later at Heathrow near London there were planes landing and taking off every 30 seconds?

I might not have been able to become a pilot or even a plane spotter, but for some strange and unexplained reason, I have always loved flying and in fact, as years went by, made a point of taking to the skies in as many different aircraft as possible.

I realise for many my desire to fly in as many different aircraft as possible might sound a bit geeky.

The other event which captivated me and made me more determined to become a fighter pilot was one day watching a local aerobatic display by the Rhodesian air force.

Thousands like me looked up in awe at these jets as the pilots performed their tricks and formation flying. How could a plane fly upside down? How could a plane do a 360 degree roll or even shoot straight up into the skies and quickly disappear?

I wasn't to know that in a few months seeing contrails or vapour trails in the skies and watching aircraft and even dreaming of becoming a pilot would suddenly disappear like the sight of vapour trails.

Chapter 2

At this stage, I need to thank the Rhodesian air force.

They were the people and planes which lead to a lifetime love of aviation.

Let me highlight some of the aircraft as well as a bit of Rhodesian air force history.

In 1949 the authorities in Salisbury Rhodesia now Harari Zimbabwe decided to build their first air Force base on a former farm called Kentucky.

The base became known as New Sarum, and later became the main air force base. It was the Roman name given to Salisbury in Wiltshire, England.

Two years later, nine Spitfires out of eleven landed at Salisbury, but these were retired and replaced 3 years later by four Vampire FB9 aircraft with 6 more arriving the following year and still more to follow.

In late October 1954 the air force took delivery of its second Dakota aircraft. That same year Four Percival Provost aircraft arrived on 4 November 1954.

The following year in 1955 another eight Provost T52 aircraft arrived.

By this stage, the fleet now included 16 Vampire aircraft FB9s, 16 T11s Provosts, 2 Dakotas and seven Pembroke aircraft as well.

In late March 1959 the first four English Electric Canberra aircraft arrived from England. From then on Canberra's arrived at the rate of two per month for four months.

John CT Miller

I don't know which countries tried rain making experiments, but in March 1961 saw Number 4 Squadron carrying out rain making experiments by releasing a mixture of salt and sand into clouds. It obviously didn't work or else sand and salt missions would have been in high demand.

In 1962, the first of 4 12 Hawker Hunter aircraft arrived to replace the Vampire of Number 1 Squadron.

This iconic British jet fighter was the first plane to fly at over 1100 KPH, and was more than 300 kilometres faster than the Vampire.

1967 saw the arrival of the first Lockheed Aeromachi AL60 B "Trojan" aircraft. Assembly began at the beginning of August and towards the end of that month 9 aircraft were completed.

It was also round about this time, plastic models of aircraft began to appear. I can't remember, but I think they came already assembled or if not, you had to glue the different parts together.

Shortly after this smaller metal aircraft models began to appear. This followed after the popularity of Dinky and Matchbox cars which were already part of some children's toy boxes.

Compared to model aircraft, cars had wheels and you could play with them on different surfaces and even on sand.

I think many like me refused to put our model aircraft in the toy box but placed them on a book shelf or attached them to the ceiling for all to see and admire.

Decades later radio controlled helicopters appeared followed by the ever present and popular radio controlled and later app driven drones.

One day back in 1961, my life was to change for ever.

When returning from junior school that day, I was blinded after a shotgun accident and my dreams of becoming a pilot crashed not just in mid-air, but disintegrated on the ground.

After the accident, my first flight was in a tiger moth, and later

Travelling by sound and smell

in one of the Rhodesian Airforce Allouette III helicopters.

Thanks to my late mom, she also managed to organise with the air force, which allowed me to sit in the cockpits of the Vampire, the Canberra and the Hawker Hunter.

Chapter 3

Within months of going blind my lifelong association with civilian aircraft began and many years later I became an aviation journalist, but much more about that later.

The nearest school for blind and disabled children was in Bulawayo some 300 miles from Salisbury.

At the time my dad and grandmother both worked for Air Rhodesia, which meant they didn't have to pay for my flights between the 2 cities.

After going blind, my first 3 years were spent flying between Salisbury now Harari airport and Bulawayo in the former Rhodesia now Zimbabwe.

Depending on the flight schedule, most times it was on a Viscount, but every now and again, the propeller driven Dakota (DC 3) the workhorse of the skies was used.

This plane flew from Salisbury via Livingston to Bulawayo.

The Dakota first took to the skies during the second world war, and believe it or not, there are still some countries some 80 years later where this plane is still being used.

The noisy Dakota cruised at about 10000 feet while the more modern and quieter Viscount was able and flew at 20000 feet.

Don't forget as you make your way around at home or out on the streets, there is only 12500 feet of oxygenated air above you. After that, oxygen rapidly declines. At 14000 feet, there is less than 43% of oxygen compared to that at sea level.

Looking back now, it is hard to believe the DC 3 was the plane that changed the way the world flew, making commercial travel

Travelling by sound and smell

popular and airline profits possible.

Compare the DC 3 cruising altitude of 10000 feet and speed of about 180 mph to Concorde with its speed of Mach 2.02 or 1520 mph and a cruising altitude of 60000 feet?

Another comparison is the number of passengers then and later.

The DC 3 could carry between 21 and 32 passengers depending on the way the seats were configured.

Who would have thought let alone dreamed one day a plane could carry up to 800 passengers?

Well, this is the case of the iconic Airbus A380.

For the next 3 years I flew regularly between Salisbury and Bulawayo on my way to school and back.

I am sure my late mom accompanied me on that first trip to Bulawayo. Apart from meeting the teachers, we also made a trip out to Matobo hills where the founder and British imperialist Cecil John Rhodes was buried.

No doubt he like many others after him considered Rhodesia to be God's own country.

In the years that followed, I flew regularly between the 2 cities with Air Rhodesia. I was certainly a frequent flyer even though frequent flyer programmes and rewards took years before such incentives were introduced to the flying public.

Apart from long weekends and the school holidays I often flew home for special occasions.

One of those special occasions was when Cliff Richard and his band the Shadows played in Salisbury.

I flew home on the Friday evening, and the following day went to the afternoon performance.

Thanks to Dyllis Stephens the lead singer for the Rhodesian group the Cyclones I got to meet Cliff and the Shadows.

Another weekend flight was to see Mickie Most and his band

play in Salisbury.

My Mom had somehow managed to contact Mickie and I received a personal invite from him.

At the gig, I also met his wife Christina, who sat with me during his show.

She invited me back when they next played in Salisbury.

Someone else who appeared with Mickie that night was Jackie Frisco, his sister-in-law.

She went on to have a couple of hits while the family lived in South Africa.

Mickie was known in South Africa as the king of rock. After leaving the country and returning to the UK, Mickie went on to become one of the world's top producers.

He discovered and recorded such acts as the Animals, Herman's Hermits, Donovan, Lulu, Suzi Quatro and Hot Chocolate to mention a few.

During my many flights between Salisbury and Bulawayo, I got to know many of the "air hostesses" as they were known then.

In those days, cabin crew were exclusively young women.

Some of these air hostesses who I became friends with, made sure whenever I flew they were working on the same flight.

The nearest high school after Bulawayo for blind children was in South Africa.

Once again on that first visit my mom accompanied me to the School for the Blind in Worcester in the Cape.

That first flight to South Africa we flew on one of the viscounts to Johannesburg from Salisbury, then boarded a B727 to Cape Town.

This was my first trip on the B727 let alone any Boeing aircraft.

Even when I attended school in the Cape, I would let my parents know the dates of the upcoming holidays and they would let my favourite air hostesses know which flight I would be on.

Travelling by sound and smell

Each time I boarded the Salisbury bound Air Rhodesia viscount from Johannesburg they were always there to meet me, and in a strange kind of way it meant freedom.

Firstly, those flights from Johannesburg to Salisbury were not just about freedom, but more importantly I got to fly again, and secondly I would be greeted by my favourite air hostesses.

There were 2 in particular, Meg and Angela. I wonder what happened to them?

In fact as a 15 year old teenager, during school holidays I was often invited out by a couple of the cabin crew.

We went to movies, picnics, boating on various lakes and they often visited me at hour family home.

After that first flight between Johannesburg and Cape Town, I had to wait years before flying the same route again.

The school insisted on all children from and around Johannesburg meeting at the main railway station to begin the journey to Worcester.

The train trip was always accompanied by at least 2 teachers from the school to make sure we all behaved ourselves. Probably more important the boys and girls remained constantly separated.

At the time, I was one of the few school children who had ever been inside an aircraft, let alone fly in one.

I was also the only child at that stage who played a steel string guitar.

This lead to me accompanying various singers on the main stage in the school hall.

However, the pupils did not seem to be interested in planes or what it was like to fly.

My ability to play guitar was far more important to them.

Thinking about it now, how do you describe to anyone let alone a blind person what it is like to fly in a plane?

13

John CT Miller

Apart from take-off, and landing, once cruising altitude is reach, not much movement is experienced.

You can try and describe to someone what it is like to fly into and through clouds and arrive on top of them and find sunshine.

You can also try and describe how small people and objects look the higher above ground the plane gets, but unless you have actually taken that first flight, your words are hard to visualise.

At the school, I was one of only 5 English speaking pupils, and also the only child who spoke out against the apartheid government.

Not only were most of the pupils racists, but worse still, most teachers were as well.

In fact, I was once beaten up by a teacher after calling him a racist pig and other choice names.

I didn't realise he was within earshot.

To say I hated my time at the school would be an understatement. The culture, the food, the Christian hypocrisy and the blatant racism were just some of the things I could not get used to.

Something else which began to happen during my school days in South Africa was the relationship with my family began to drift apart.

Unlike my time at the school in Bulawayo, I only flew home twice a year.

The South African school holidays were also at different times of the year compared to those in Rhodesia.

This meant when I went home midyear, Rhodesian children were not on holiday.

The only time I got to spend time with other school children and my brothers was during the December summer holidays.

But as the years went by, it became harder to catch up with family matters.

By the time I matriculated and left the school for the last time

Travelling by sound and smell

before returning to Salisbury there was only one other pupil who had actually travelled in an aircraft.

I couldn't wait each time to board the Air Rhodesia Viscount from Johannesburg to Salisbury.

Every time I arrived at the airport in Johannesburg and walked through the door of the waiting Viscount, the sense of freedom and home-coming returned.

When I talk about home-coming, this had to do with once again being a passenger, and nothing to do with returning home to my family.

It would be a couple of years before my next and last flight on the Viscount after I gave up my job as a personnel consultant and flew back to Johannesburg to look for another job.

For almost the next 2 decades my association with South African Airways and Boeing aircraft began.

In the years that followed, I was also able to fly on the Junker JU 52, the Dakota (DC 3 and DC 4) through to the Concorde and many other aircraft in between.

The ever increasing choice of planes took me much later to many airports in Southern Africa, Europe, America, the Middle East and Asia.

In fact, in the 90s I actually flew from Gatwick London to Boston in the US in a Boeing 747-200 and after a 2 hour stopover flew back to Gatwick.

This trip came about at the invitation of Richard Branson, who I interviewed at his home, and who asked me if I had ever flown Virgin Atlantic?

When I said I had not flown on any Virgin Atlantic planes, he told me his airline had a flight leaving the next day to Boston. He asked if I would like to experience Virgin Atlantic.

At that stage, Virgin had not begun flights to South Africa.

John CT Miller

It goes without saying I immediately accepted his offer and departed from Gatwick the next morning.

My trip gave me the opportunity to witness what or how the cabin crew dealt with passengers.

I remember there was one Australian passenger, who no matter what the crew did, he was never satisfied: complaining about everything.

The patience and professionalism they displayed was incredible.

I bet if you spoke to flight attendants around the world, they will most likely tell you of many similar tales.

Before encountering some headwinds, let me go back in time.

Becoming an international member of the cabin crew was for many most likely the dreams of all those who joined airlines.

I am sure in those early days many young women and later men were attracted by the thoughts of "seeing the world", and not having to pay for it.

However, once these women passed the various tests including height, weight and no doubt looks, the successful applicants were first assigned to flying domestic routes.

If you proved competent and successful on domestic routes, you were invited to join the cabin crew on regional routes.

After time on these routes, the air hostess might be lucky enough to be selected to become part of the cabin crew on international flights.

As part of joining the international cabin crew meant you had reached the top of the ladder so to speak.

I guess it is a bit like passengers who turn right or left on entering the aircraft. Turning left meant most times you had made it and were able to afford flying business class.

For many years in the past, international crews were given between 5 and 8 days to relax on arrival in capitals around the

Travelling by sound and smell

world. They also didn't have to pay for hotel accommodation.

These air hostesses were extremely popular and sort after by many. Family and friends would look forward to their daughter and friend returning from some overseas trip with not just various duty free items, but also treats and trinkets from foreign fields.

Depending on the airport and their relationship with those working at immigration, often they were not asked if they had "anything to declare?"

Family and friends couldn't wait for them to return to find out what treasures they had brought back with them.

It wouldn't just be perfumes and makeup, but such simple items as not available or unusual food delicacies.

The first air hostess to take to the skies was in 1930 in America, followed 4 years later in Europe by Swiss Air and later now known as British Airways.

The sexism previously experienced by flight attendants as they are now known is mostly a thing of the past. Previously these women had to quit flying aged 30, and it was only in 1993 they didn't have to regularly jump on a scale. They also couldn't be married or have children.

These pioneering women also made sure men could join them in the air and were entitled to the same rights and also made sure people of colour were given the chance to fly.

Contrary to what many passengers think, flight attendants are not just there to serve food and refreshments.

Just like pilots, they have to undergo regular inflight training. This includes medical and safety training, and responding to emergencies.

Each year about 44 thousand passengers need medical assistance.

Many airlines insist on flight attendants returning every 18 months to get requalified.

They have to be able to evacuate a plane carrying over 300 passengers within 90 seconds.

On average about 30 full plane evacuations take place each year out of 40 million flights.

The biggest danger to passengers is inflight turbulence.

Today climate change is leading to much more turbulence.

However with technology advancing, pilots are able to fly around storms. They are also able to track and watch volcanoes, track earthquakes and check on aircraft around them and near airports.

Head of flight attendant training for Delta surprisingly used to be a musical theatre performer.

Chapter 4

In days gone by, there were also no things like air bridges. Passengers walked to the terminal after a mobile stairway was rolled out and rapidly attached to the plane's door.

Depending on the arrival time and the number of aircraft near the terminal, passengers were often taken by bus to the terminal.

While I know many enjoy petrichor, but there is nowhere better than in Africa to enjoy the smell of those first rains falling on baked earth.

But for me even better was the smell of aviation kerosene. It was just as invigorating and to this day, one of my favourite smells.

Sadly when boarding an aircraft and with the use of air bridges, kerosene these days is seldom smelt.

Back in the good or bad old days, smoking was also allowed on board aircraft. However by the 90s, lighting up a cigarette on aircraft had been stopped and banned.

Thinking back to those days, I don't know who ever thought a flimsy curtain separating the smoking cabin to the non-smoking section made very much difference? I guess it worked reasonably well on a short hall flight, but wondered how successful the curtain was on long distance flights?

Who remembers the flip up ashtray in the armrest?

Later I was often reminded of those smoked-filled cabins at some airports around the world which offered passengers a dedicated smoking area.

The second you opened the door and entered one of these areas, you were greeted and enveloped by a cloud and wall of smoke.

John CT Miller

Those were also the days even if you boarded a one hour lunchtime flight certain airlines offered a full meal and drinks service.

The minute the aircraft reached cruising altitude, the air hostesses rushed out with their trollies and began serving meals. As the meals were handed to passengers, a drinks trolley followed.

Those days are long gone. Today passengers are lucky enough if they are given a pre-packaged sandwich or slice of cake, but most times it would be a tiny packet of peanuts or corn bites.

In fact with the advent of low-cost airlines, passengers now have to pay for any snacks. Some even threatened to charge passengers to use the toilet.

Many of these low-cost airlines also charge passengers for their hand luggage. It goes without saying there was also an additional charge if you wanted your suitcase stored in the cargo hold.

Another tradition which has disappeared is handing out a boiled sweet to each of the passengers before take-off and another sweet when descent began.

The idea was by sucking on the sweet, it would prevent blocked ears.

I do wish they gave babies and very young children something to suck on. This would certainly help them from any discomfort.

Chapter 5

After working in South Africa for a couple of years, I decided it was time to move on and look for a job in London.

My favourite band the Beatles had already split up and I had missed out on the swinging 60s, but London still had its appeal.

The swinging 60s had come and gone, but streets like Carnaby street, which became the place to be seen for fashion and lifestyle retailers was still there.

My parents were originally born in the UK before settling in Rhodesia.

At that time Rhodesia was like a mini Britain. In Rhodesia unlike in its neighbour south Africa many English radio programmes were heard on the local airwaves, and many English products were available in most shops.

My family had grown up playing monopoly with such famous places to buy as Mayfair, Park Lane along with various well known railway stations.

Even though at that stage I had not visited London, many of the street names and areas thanks to monopoly had a familiar sound to them.

I went to speak to a couple of travel agents, and one told me about a deal to London which included a few stop-overs before reaching the British Capital.

This sounded great. It also meant perhaps I could fly on other aircraft not seen in Southern Africa.

At the time, the fact I was also going to visit some other European capitals for the first time was not that important to me.

John CT Miller

When I told the travel agent I was more interested in the types of aircraft used on the different routes, I am sure he thought I was mad. Why wouldn't I be excited about visiting some of the great European capitals of the world?

Thanks to the travel agent he managed to devise a schedule based not on cities, but on how many different aircraft I could fly in.

This was going to be my first overseas trip and flying on the B707.

After arriving in Europe, I remember stopping off for a couple of days in Amsterdam, Athens, Milan and Vienna.

The planes I got to fly in included the Trident, the DC-9, the Caravelle, the BAC-111 and the A300.

At the time, I had no idea how in later years the Airbus fleet of aircraft would become my sort after manufacturer.

While in Vienna, I visited the nearby woods and sampled various wines from several small wine estates.

Making our way through the woods was certainly a new way of sampling individual wines.

I couldn't travel all that way to Vienna, without finding a restaurant serving traditional Wiener Schnitzel together with some redcurrant jelly, and a potato salad.

There can't be many cities which have given their name to a now famous dish across the world.

The Wiener Schnitzel is one of the most famous dishes in Vienna, and was traditionally made using veal.

However, today the Schnitzel is made using chicken, pork, and even turkey meat.

Another dish I remember trying was the Sachertorte.

The cake has a thin layer of apricot jam in the middle and coated in dark chocolate icing. The cake is originally served with whipped cream.

Travelling by sound and smell

This cake is so famous in the country, there is even a National day dedicated to it.

Another dish I had to try was their Apfelstrudel.

The Apfelstrudel is one of the best-known pastries in the country. It features a super thin and crisp pastry, which is much like filo. The filling is rolled inside, and the dish is traditionally served with the swirls showing off using the pastry, along with apples. The filling inside the Apfelstrudel is made with sugar, cooking apples, raisins, cinnamon, and bread crumbs.

Sadly in the UK today the pastry used tends to be stodgy and heavy. This is typical of most pies in the UK.

During that trip I also visited various attractions in Athens and Amsterdam.

Once in London , my visit to find a job was soon abandoned. I was lured not by the lights of London but by the sounds of nightclubs and bars.

Before I knew it, my money was about to run out. Fortunately, I had kept my return ticket.

On arrival back in Johannesburg I was greeted by much laughter and jokes at my expense.

Oh well, for the next few years it was back to flying domestic routes on the B737 or the B727 to various cities in South Africa.

Chapter 6

A few years later I was once again London bound, and this time decided to stop over and visit my parents and my sister, who were still living in Salisbury.

London this time, there would be no bars and nightclubs. I was there to begin a 3 year course in physiotherapy.

Talking to a travel agent, I discovered I had a choice to fly with Air Rhodesia in their recently acquired 3 Boeing 720 fleet or South African Airways fleet of B737 aircraft.

The choice was simple.

SAA had decided not to buy the older B720 and this might be my only chance to fly in it.

At the time, domestic routes in South Africa were served by the B737 or the B727.

Before visiting my parents, I had already flown on both the B727 and the B737.

While back in South Africa after my last trip to London, I had formed my own band, and when I decided to leave again, sold all my equipment including my much treasured Gibson electric guitar.

What a fool I was.

I had not seen my family for some years, and thought it was a good idea to spend 10 days with them.

One day while at the family home I heard what I thought was the sound of fire crackers in the distance.

I found this strange especially as it was in the middle of the day.

I asked my sister "why are people letting off fireworks?"

Travelling by sound and smell

Casually she replied it wasn't fireworks, but terrorists using their ak-47 rifles.

That did it for me. I couldn't wait to get out the country, and began counting the days before my flight to London resumed.

I have always enjoyed take-offs and the climb to cruising altitude. But on that day, I couldn't wait and even possibly held my breath at times until the SAA B747-SP reached cruising altitude away from any terrorist missiles.

Once we reached cruising altitude and were out of Rhodesian air space I was able to relax again and enjoy the rest of the flight.

Even before my visit, I had been uneasy about a stop-over in Salisbury.

What made this trip to Salisbury even more worrying was the year before I heard the dreadful news on September 3 1978 an Air Rhodesia Flight 825 had crashed, and the following year in February another Viscount was shot down.

Chapter 7

Flight Air Rhodesia 825 was one of my much loved Viscount aircraft called Hunyani which had been shot down by terrorists shortly after take-off from Kariba. It was named after a river which flowed from Kariba down to Salisbury.

That Viscount and that river each held fond memories for me.

I had flown on this aircraft many times in the past.

The Hunyani river flowed a few miles from our house.

As kids my brothers and I would often go down to the river to fish and always on the lookout for hippos or crocodiles.

The ill-fated flight from Kariba that Sunday afternoon took off with 52 passengers and 4 flight crew.

Within 45 minutes 48 were dead. 38 were killed instantly when the plane crashed into the surrounding bush and 10 other passengers were killed in the following cold-blooded massacre at the site.

Air Rhodesia Flight 825 was a scheduled passenger flight shot down by the Zimbabwe People's Revolutionary Army (ZIPRA).

About 5 minutes after the aircraft took off, a Soviet-made heat seeking Strela-2 surface-to-air infrared homing missile hit the starboard wing and minutes later after a mayday call the plane crashed into a nearby cotton field exploding on impact and instantly killing 38 passengers.

Of the 56 people on board, 38 died instantly. Eighteen survived, albeit with injuries, but were able to climb out of the wreckage.

After briefly attending to the other survivors, one of the passengers led four others including 2 young newlyweds and a

Travelling by sound and smell

mother and her four-year-old daughter off in the direction of a nearby village in search of water.

Meanwhile, the other 13 remained close to the wreckage. As the survivors waited, nine guerrillas made their way towards the crash site, and reached it about 40 minutes later.

Seeing figures approaching, three of the 13 survivors remaining at the crash site decided to hide and took cover in the surrounding bush.

The 10 passengers including 4 women and 2 girls aged 11 and 4 years old continued waiting near the plane.

When the terrorists armed with AK-47 guns arrived, they told the survivors saying they would summon help and bring them water.

They spoke in English, both to the survivors and among themselves. They told the passengers to gather around a point a few metres from the wreckage.

When the survivors said that some of them were too badly injured to walk, the terrorists told the able-bodied men to carry the others.

Once the passengers were assembled into an area of about 10 square metres, the terrorists who were standing about 50 metres away now raised their weapons.

Before they opened fire they shouted "You have taken our land,"

"Please don't shoot us!" one of the passengers cried. Those that survived the initial bursts were bayoneted including a mother and her 3-week-old baby.

Meanwhile the others who had collected water from the nearby village were almost back when they suddenly heard the gunfire.

Thinking it was personal ammunition in the luggage exploding in the heat, they continued walking towards the crash site, calling out to the other passengers, who they thought were still alive.

This alerted the terrorists that there were more survivors. One of the guerrillas told the group to "come here".

The terrorists then opened fire in the general direction, which allowed the remaining survivors to flee.

While some of the survivors were hiding behind a nearby ridge, they saw the terrorists return about 2 hours later to the crash site, and witnessed the terrorists looting the wrecked cabin and some of the suitcases strewn around the site. The terrorists filled their arms with passengers' belongings, before leaving.

The survivors were found over the following days by the Rhodesian Army and police.

The next day after the terrorist attack, the ZIPRA leader Joshua Nkomo publicly claimed responsibility for shooting down the Hunyani in an interview with the

BBC's Today programme, saying the aircraft had been used for military purposes, but denied that his men had killed survivors on the ground.

Five days later, the Anglican Cathedral of St Mary and All Saints in Salisbury held a memorial service for the passengers and crew of Flight 825.

About 2,000 people crowded inside the cathedral, with another 500 standing outside on the steps and pavement: many citizens across the country listened to the service at homes on portable radios.

Dean John de Costa gave a sermon criticising what he described as a "deafening silence" from overseas. "Nobody who holds sacred the dignity of human life can be anything but sickened at the events attending the Viscount", he said. "But are we deafened with the voice of protest from nations who call themselves civilised? We are not! Like men in the story of the Good Samaritan, they pass by on the other side.

Travelling by sound and smell

The ghastliness of this ill-fated flight from Kariba will be burnt upon our memories for years to come. For others, far from our borders, it is an intellectual matter, not one which affects them deeply. Here is the tragedy!"

Meanwhile, Air Rhodesia began developing anti-Strela shielding for its Viscounts. But before this work was completed, ZIPRA shot down a second Viscount, Air Rhodesia Flight 827, on 12 February 1979.

This time there were no survivors.

Following the second terrorist attack, Air Rhodesia created a system whereby the underside of the Viscounts were coated with low-radiation paint, with the exhaust pipes concurrently shrouded.

According to tests conducted by the Air Force, a Viscount treated could not be detected by the Strela's targeting system once it was over 2,000 feet (610 m). There were no further Viscount disasters in Rhodesia.

Unbeknown to civilian passengers and most Rhodesians in September 1978, there had been 20 reported attempts to shoot down Rhodesian military aircraft using these weapons: none of which had been successful.

Some Rhodesian Air Force Dakotas had been hit, but all had survived and landed safely. No civilian aircraft had yet been targeted during the Bush War.

Over 30 years later in 2012, and inaugurated on 1 September that year A memorial to the victims of the two Rhodesian Viscount incidents, dubbed the Viscount Memorial, was erected on the grounds of the Voortrekker Monument in Pretoria, South Africa.

The names of the dead passengers and crew are engraved on two granite slabs that stand upright, side by side, the pair topped by an emblem symbolising an aircraft. A pole beside the monument flies the Rhodesian flag.

John CT Miller

A British parliamentary motion put forward by Labour MP Kate Hoey in February 2013 to retrospectively condemn the Viscount attacks and memorialise the victims on the anniversary of the second shootdown prompted outcry in the Zimbabwean press, with the government controlled Herald branding it 'racist".

Before I forget, that memorial service can still be found and listened to on YouTube.

Chapter 8

After some time in London I bought a top of the range Marantz hi-fi with 2 large speakers and a big tuner amplifier.

At the time, I did not think about what to do with it if and when I returned. I certainly wasn't going to leave it behind in London.

Six months later, I began to make my plans to return to South Africa.

When I got to Heathrow that evening with suitcase and hi-fi in hand, I knew my luggage would be far over the weight limit.

Arriving at the check in desk, I wasn't sure what to do. I certainly didn't have money to pay the excess baggage charges.

While waiting in line, I still had no idea what to do.

Before they could weigh my luggage, I casually asked the staff if they had heard from the South African embassy?

Sounding a little surprised, the check-in clerk said "no."

I knew at that time of the day the embassy would be closed, and there was no way of them checking up on my made-up story.

Pretending to be somewhat indignant and surprised, I told them the embassy was supposed to have spoken to SAA and cleared the hi-fi and speakers.

One thing I had learnt was most times when one quotes officialdom, people are not sure what to do.

It was certainly the case that day.

The words "South African embassy" did the trick and they immediately allowed me to board with my very expensive hi-fi.

One hurdle overcome, but what to do when I reached the airport in Johannesburg?

John CT Miller

During that overnight flight on the B747 I got to talking to one of the air hostesses, and explained my predicament.

She must have taken pity on me and offered to bring the hi-fi through as part of her luggage.

I guess in those days home grown crew were probably treated differently.

Once I had cleared customs, I waited for my new found friend to arrive with my hi-fi, which she did some minutes later.

This Marantz hi-fi was their latest model and was not available in South Africa.

My return this time wasn't greeted with delight, but with recriminations by my parents.

They told me I had blown my last chance of getting a "decent job".

Oh well, it was back to looking for a job once again as a telephone operator.

One month later I managed to get a job at South African Associated Newspapers answering their switchboard.

Chapter 9

The switchboard job didn't last long before I was asked by the news editor of the Rand Daily Mail if I ever thought about becoming a journalist?

Naturally I said Yes, but added the management might have some reservations about employing a blind person as a journalist.

In the preceding years I had discovered how "sighted" people especially companies and women discriminated against disable people.

Funny enough, the discrimination didn't come from black people, but white people.

Finally some weeks later after management had indeed tried to find every excuse in the book to prevent me from becoming a journalist I began their cadet course.

The cadet course was held in Port Elizabeth and better still after settling in, I discovered the city had a flying club.

This was great news. Not only was there the possibility of doing a few flights, but also the drinks at the clubhouse were a lot cheaper than city pubs and bars.

It wasn't long before I was made an honorary member of the flying club.

Admittedly, I would not be flying on any Boeing aircraft, but smaller planes. At least I would be able to take to the air again.

Some 3 months later I returned to Johannesburg and began work as a journalist on the Rand Daily Mail and with it soon began flying on the B737 aircraft again.

Five years later in 1985, the newspaper closed down and soon after that I was back to my next overseas trip on the B747-SP to Milan.

The only reason for choosing Milan was because some friends Tamara and Paola and their family had moved near to that city some months earlier.

During my time on the newspaper I had also become the Southern African correspondent for the bible of the entertainment industry: Billboard.

After visiting a couple of European capitals on various trains, I flew to London on a B737.

Billboard's offices at the time were close to Carnaby street. While at the office I popped down to revisit this once famous street.

It was disappointing to discover the street now housed cheap and tacky. I doubt if any of the rich and famous would want to be seen within 50 metres of that street.

My trip to Europe was to try and sell South African music to various overseas labels.

Looking back now, it was the height of apartheid and South Africa was definitely not the flavour or sounds of the month.

Thanks to apartheid, I was not able to sell any of the various South African independent artists to any of the overseas labels.

How naive I was.

Chapter 10

Arriving back in Johannesburg I soon joined the Citizen newspaper and for the next 4 years worked as a crime and court reporter.

In one of my previous chapters I told you about the shooting down of the 2 Air Rhodesia Viscounts by terrorists, but 8 years later there was another significant aircraft disaster this time much closer to home so to speak.

In 1987, Gordon the late editor of the entertainment pages of the Citizen newspaper and I went to watch South Africa play Australia in cricket at the Wonderers in Johannesburg.

Half way through that afternoon an announcement came over the public address system.

The announcer said a SAA plane had crashed into the Indian Ocean and there were no survivors.

For a few minutes, there was a stunned silence around the stadium before the match continued.

On 28 November 1987 Flight 295 South African Airways Boeing 747-200 named Helderberg crashed into the Indian ocean 248 km northeast of Mauritius killing all 159 passengers and crew.

The plane had earlier left Chiang Kai-shek International Airport in Taipei Taiwan destined for Johannesburg via Mauritius.

Within 2 days of the search beginning parts of the wreckage was found.

Days after the crash, an extensive salvage operation was launched to try to recover the aircraft's flight recorders, one of which was later recovered from a depth of 4,900 metres (16,100 ft).

At some point during the flight, believed to be during the beginning of its landing approach, a fire started in the cargo section on the main deck which was probably not extinguished before impact.

The United States Navy sent aircraft from Diego Garcia, which were used to conduct immediate search and rescue operations in conjunction with the French Navy.

Later, South Africa sent a total of six naval and civilian vessels to the search area.

The South Africans mounted an underwater search, named Operation Resolve, to try and find more wreckage.

After recovering much of the wreckage from 4,000 m (13,000 ft) below the ocean's surface, the aircraft's fuselage and cabin interior were partly reassembled in one of SAA's hangars at Jan Smuts Airport where it was examined and finally opened for viewing to the airline's staff and selected members of the public.

The underwater locator beacons (ULBs) attached to the flight recorders were not designed for deep ocean use. A two-month-long sonar search for them was carried out before the search was abandoned on 8 January 1988 when the ULBs were known to have stopped transmitting.

A couple of days earlier on 6 January 1989, the cockpit voice recorder (CVR) was salvaged successfully from a record depth of 4,900 metres (16,100 ft) by a remotely operated vehicle but the flight data recorder was never found.

An official commission of inquiry was later chaired by Judge Cecil Margo, and found that inadequate fire detection and suppression facilities in the class B cargo bays were the primary cause of the aircraft's loss.

Cecil Margo was known to be a supporter of SAA.

For many, the commission ignored the most important question:

what was the source of the fire and who had been responsible for loading the aircraft?

This lack of a conclusion led to many conspiracy theories and a subsequent post-apartheid investigation followed years later. It too was also shrouded in secrecy.

The crash was the first fire incident on the 747 Combi and one of few fires on wide-body aircraft.

In January 1992, the journal of the Royal Aeronautical Society (RAeS) reported that the inquiry into the in-flight fire that destroyed SAA Flight 295 might be reopened because the airline had allegedly confirmed that its passenger jets had carried cargo for Armscor, a South African arms agency.

The inquiry was not reopened, and this led to more speculation and more conspiracy theories about the nature of the cargo that caused the fire.

Examples of such theories include: The SADF was smuggling the hoax substance red mercury on the flight for its atomic bomb project.

Reports from the Project Coast investigation suggested there was a waybill showing that 300 grams of activated carbon had been placed on board the plane.

Dr David Klatzow an internationally forensic scientist, was retained to work on the case by Boeing's legal counsel around the time of the official enquiry.

He then criticised the Margo commission for spending an inordinate amount of time looking into "relatively irrelevant issues" and that the commission ignored the most important question: what was the source of the fire and who had been responsible for loading it onto the aircraft?

Klatzow believed there were certain irregularities in parts of the commission transcript that indicated parts of the cockpit voice recorder transcript had been concealed.

Klatzow suggested the fire likely involved substances that would not normally be carried on a passenger aircraft, and the fire was not likely caused by wood, cardboard, or plastic materials.

At the time, South Africa was under an arms embargo and the government had to buy arms clandestinely.

His theory suggested the South African government placed a rocket system in the cargo hold, and that vibration caused the ignition of unstable ammonium perchlorate, which is a chemical compound used as a missile propellant.

In 1996, the Truth and Reconciliation Commission set up by the post-apartheid African National Congress South African Government, investigated apartheid era atrocities.

Unlike most other hearings of the TRC, the hearing into SA 295 was conducted in camera,

and without any representation from the South African Civil Aviation Authority.

Judge Margo was also not summoned to answer any of the allegations made against him.

In the early 90s, I got to know and befriended someone high up in SAA. I often visited him and his wife over weekends.

During one of my visits, one night after enjoying a braai or barbeque, he said that one day the truth would come out about the crash.

He would not tell me more, but I am sure to this day that plane was carrying something it shouldn't have been carrying.

It is also true the recorded control tower tapes and certain documents in Thai pay mysteriously went missing.

Many people believed if SAA admitted to the illegal cargo, the compensation claims would bankrupt the airline.

Travelling by sound and smell

Unbeknown to me and many others years later SAA was bankrupted, but not from crash compensation, but after years of mismanagement, ANC cadre deployment and theft of millions from its accounts by the African National Congress.

In just over 10 years more than r50 billion was given to SAA to keep it in the air.

The airline in many ways became the private airline of the corrupt ANC. Members of the ANC, their friends and families would fly without paying for their flights.

Chapter 11

In the following few years I continued flying on the B737 often to Durban to visit a musician friend of mine Kevin Mason.

I normally left work on Friday and returned on Sunday evening.

During my years flying I have only missed 3 flights.

I should add, all through my own fault.

The first happened when I was due to catch a flight from Durban back to Johannesburg one Sunday night.

Waiting around for one hour to board the flight was not for me, so I decided to leave it to the last minute to check in.

Unfortunately for me, my last minute timing did not work.

By the time I arrived at the airport, and went to check in, I was told the gates had already closed. They were not prepared to open and allow me to board.

The next flight was only the following morning. This would be too late. I had to be at work at 7am.

Kevin my friend offered to drive through the night to get me back to Johannesburg as long as I paid for the petrol.

Six hours later we arrived back in Johannesburg.

The next time I missed my flight was some years later.

I had arrived in Johannesburg from London on an Emirates B777 and was due to get the connecting flight to Durban 2 hours later.

One of the ground crew at OR Tambo airport in Johannesburg met me off the plane. After clearing customs and collecting my luggage, I decided there was time to stop and get a good old South African burger.

Travelling by sound and smell

The member of staff was really great and she decided to stay with me until it was time to board the next flight.

We both ordered a burger and something to drink and completely forgot about the time.

When she next checked her watch, I had missed the flight.

A friend of my brother was going to meet me at the airport in Durban.

Well, it goes without saying the friend got there on time, and waited and waited but no John CT Miller appeared.

I did not have the friend's cell phone number and contacted my brother telling him the inbound flight from London had been delayed.

My brother also did not have her number and had to wait until she returned home without yours truly.

Back in Johannesburg the next flight on a B737 to Durban was only 90 minutes later, which this time I made sure I boarded.

Meanwhile my brother's friend when told what had happened, had to immediately turn round and drive back to the airport.

As you can imagine, she was not very happy.

The last time I missed my flight happened in Manila the Philippines.

Once again, I decided I did not want to wait the 3 hours at the airport before boarding the flight with Thai airways.

I don't know why, but that day I also somehow got the time wrong. I decided 20.00 hours was 10pm, which of course you all know 20.00 hours is 8pm.

Naturally by the time I arrived at the airport, the check-in desk was long closed.

Eventually after finding the Thai airways offices somewhere in the building, I made my way up to speak to them.

My excuse this time was traffic congestion in Manila.

41

John CT Miller

For those who have visited Manila, you will know traffic in that city is a nightmare.

The Thai Airways rep was not particularly understanding nor very helpful or friendly, and reminded me Manila traffic was always congested. It was up to me to make sure I got to the airport on time.

She then insisted I paid an extra £400 for the new flight. After pleading poverty she finally reduced it to £150, and booked me on the next flight the following day.

However, according to her, the connecting flight from Bangkok to London was fully booked and I would have to spend an extra day in that city.

Beggars can't be choosers, so I begrudgingly handed over the £150 and returned the next morning.

I didn't believe the connecting flight was fully booked, so when I arrived in Bangkok I immediately went to the help desk and fortunately dealt with a really helpful lady, who checked seat availability, and booked me on the B777 flight leaving some 90 minutes later.

So much for that rep telling me the flight was fully booked.

All's well that ends well, and I got back to London on time.

Chapter 12

Before I continue, I cannot tell you the number of the following announcements I heard while waiting to either board or seated in aircraft across the world.

These announcements were probably also made when I failed to arrive on time.

For me, one of the most annoying announcements was often heard even before boarding an aircraft.

"This is the final boarding call for passengers*** and *** booked on flight 372A to some or other destination. Please proceed to gate 3 immediately."

I never understood why these idiots needed to be called. Not once, not twice but sometimes 3 times.

These announcements were most times heard at Heathrow or Gatwick. I am not surprised as it was often English passengers, who it seemed could not fly anywhere without spending time drinking in one of the many bars or restaurants and often getting pissed before they even boarded the plane.

Surely they are at the airport to catch a flight?

Once seated on the plane and before take-off, the first announcement would come from the flight deck. "Cabin crew cross check doors."

The next announcement including a demonstration would be "We ask that you please fasten your seatbelts at this time and secure all baggage underneath your seat or in the overhead compartments. Please make sure your seats and table trays are in the upright position for take-off. Please turn off all personal electronic devices, including laptops and cell phones. Smoking is prohibited for the duration of the flight.

On behalf of the crew I ask that you please direct your attention to the monitors above as we review the emergency procedures. There are six emergency exits on this aircraft. Take a minute to locate the exit closest to you. Note that the nearest exit may be behind you. Count the number of rows to this exit. Should the cabin experience sudden pressure loss, oxygen masks will drop down from above your seat. Place the mask over your mouth and nose, like this. Pull the strap to tighten it. If you are traveling with children, make sure that your own mask is on first before helping your children. In the unlikely event of an emergency landing and evacuation, leave your carry-on items behind. Life rafts are located below your seats and emergency lighting will lead you to your closest exit and slide. We ask that you make sure that all carry-on luggage is stowed away safely during the flight. While we wait for take-off, please take a moment to review the safety data card in the seat pocket in front of you."

In the early days, these announcements were live with one of the cabin crew behind the microphone and another standing in front demonstrating what to do. However, in later years, technology took over and at the push of a button a video would play.

Once in the air, the next announcement would most often be from the flight deck. "Good afternoon passengers. This is your captain speaking. First I'd like to welcome everyone on Flight 86A. We will be cruising at an altitude of 33,000 feet at an airspeed of 400 miles per hour. The time is 1.25 pm. The weather looks good and with the tailwind on our side we are expecting to land in London approximately fifteen minutes ahead of schedule. The weather in London is clear and sunny, with a high of 25 degrees

Travelling by sound and smell

this afternoon. If the weather cooperates we should get a great view of the city as we descend. The cabin crew will be coming around in about twenty minutes time to offer you a light snack and beverage. I'll talk to you again before we reach our destination. Until then, sit back, relax and enjoy the rest of the flight."

Chapter 14

Back to matters on the ground.

After working at the Citizen newspaper for a couple of years, I got a job as a consumer reporter at The Star newspaper.

It wasn't long before realising the Star was the only newspaper which did not have an aviation reporter.

Was this my chance to possibly fly for free?

I spoke to the news editor and said I was prepared to act as an aviation reporter.

Fortunately he agreed. But said on condition it didn't interfere with my job as a consumer journalist.

I have to admit, my offer was mostly driven by self-interest.

Forget about an open sky policy by the government. My appointment meant open skies for me. Hopefully free locally and even possibly free overseas flights were waiting just for me on the runway?

I was ready for take-off.

The next few months saw me contacting PR's from local and international airlines asking for stories while at the same time cultivating contacts.

Airlines were no different to other companies, free write-ups saved them having to pay for adverts.

There was also a benefit to airlines.

It meant the airline could deal with one person and not rely on press releases sent to the newsroom, hoping their story would be used.

Travelling by sound and smell

First on the list of befriending airline press officers was SAA.

At the time SAA had the monopoly over the domestic skies.

Even though the ANC was about to take over the government, SAA was allowed to continue its monopoly.

One thing people soon learnt with the ANC government, the new players on the block, they were prepared to sign up to any international treaty in order to be seen as progressive.

However, most times this was simply window dressing.

Due to the political scene in South Africa and the previous sanctions along with a siege mentality, the South African skies remained a government affair for many years.

Whatever the state owned airline wanted became law.

As with any monopoly, customers or passengers had no choice, and for years had to put up with poor service and inflated fares on all domestic routes.

It was different with international fares. At least passengers had other options apart from SAA.

From the late eighties things began to change and by 1991 the South African government passed legislation deregulating domestic skies.

Luxavia based in Luxenberg operated between that country and South Africa had previously used a couple of quickly painted over SAA planes and in 1991 decided to offer South African citizens a choice on domestic routes.

The new domestic airline was called Flitestar.

Four brand new A320 aircraft soon took to the skies over South Africa. This was the first time a fully computerised or (fly by wire) aircraft had been seen in South Africa.

Flitestar made sure emphasis was on customer care and superior on-board service. The Captain even greeted passengers at the aircraft door when boarding.

The new airline started gaining in popularity, so much so that SAA was forced to withdraw the A300 from regular service, only returning once Flitestar was forced to close down.

Within months, Flitestar took 25% of the domestic market and were carrying a load factor of 63%.

SAA still controlled many things and embarked on a campaign of dirty tricks.

Flitestar shared the same ticketing system (SAFARI) as SAA.

When passengers tried to reserve a seat on Flitestar, the SAA agents started making flights on the shared reservation system appear to be fully booked.

The dirty tricks didn't stop there.

SAA aircraft tried to delay Flitestar departures by blocking their aircraft on push-back. Air Traffic Control always gave SAA preference.

The state owned airline also increased its commissions to travel agents and extended its frequent flyer programme to its domestic service.

It didn't take long before I was able to pick up the phone, speak to the Flitestar PR and if a seat was available, fly to Durban and Cape Town for weekends.

SAA was not that accommodating with its free flights.

Free flights with the national airline would be by invite only, and most times only if SAA needed some publicity.

The Airbus A320 was a wonderful plane and in years to come, Airbus aircraft were always my first choice, but more about Airbus later.

A couple of years after flitestar first launched, talks had begun on replacing the A320 Bahrain service with a B767 and also using this aircraft to start a service to the Seychelles and even to replace the B747 service to Luxemburg.

Travelling by sound and smell

This was too much and too far for the government and SAA, and on Monday 11th April 1994 Flitestar and Luxavia announced that they would cease operations that very day.

SAA had won, and the idea of an (open skies) policy was a myth.

Chapter 15

Even though SAA were selective about supplying stories to the press, most of the aviation reporters also kept in regular contact with the Boeing representative and press officer who was based in Johannesburg.

One of my first free flights with SAA was in the early 90s when it began operating flights over Africa. Up until then, SAA was banned from overflying almost all African countries.

That first flight was from Johannesburg to Abidjan in the Ivory Coast on a B737.

We did a refuel stop-over in Kinshasa Zaire before flying on to Abidjan for one night.

This inaugural flight was beset with problems.

The airport authorities in Kinshasa refused to accept credit card payment for the fuel.

While we all waited for some kind of resolution, we spent a couple of hours in the terminal.

The building was rundown. Broken light fixtures, paint peeling off most walls etc. The only part of the terminal which had air conditioning and looked well-kept belonged to Swiss Air.

It took a quick collection of cash from all those on board to save the day.

On our way back to Johannesburg, the on board toilet system broke down, and the toilet doors for most of the flight were locked.

Fortunately all on board had strong bladders, but this didn't stop the plane from making a rapid descent into Johannesburg.

Travelling by sound and smell

As soon as the plane landed, we all made a dash for the nearest toilets in the terminal.

The food on the flight back was also terrible.

In my article the next day I said the catering should have been left to KFC, the food would have been better.

My comments and criticism didn't go down well with SAA, and for the next few months I was persona non grata.

It took a while and after eating humble pie and not KFC did help, and all transgressions were forgiven and forgotten.

I was back on the party list again.

Another flight with SAA was far more memorable and enjoyable.

The airline invited all the aviation reporters on a weekend freebie to Mossel Bay in the western Cape.

We took off from Johannesburg on a B737 landing in the small town of George.

It is claimed the only reason for an airport in George was to accommodate the then president of the country who lived nearby.

When we landed at the airport in George we were all taken by coach to a hotel in Mossel Bay. On arrival, the PR at the hotel asked us if we liked eating oysters?

I told her "no", to which she replied "well when you leave here, you will have changed your mind."

She was right. We had wild oysters for breakfast, lunch and dinner, served sometimes with a dash of tabasco, or lemon juice and black pepper and all proved delicious.

These were different in taste to the farmed oysters, and obviously were a hell of a lot more expensive.

During the weekend getaway, we also made a trip to the lighthouse Cape St Blaise.

None of our group wanted to climb to the top, so I decided to show them what a blind person can do and made my way up to the top.

I must admit coming down was more terrifying than the climb to the top.

I obviously didn't let on my reservations and fears about climbing down.

Another excursion that weekend was to a Cape seal colony. Even though the boat was about 20 metres from the rocky outcrop, I couldn't believe the stench coming from the small island.

Believe me, you don't have to be up and close to find out for yourself.

In those days, some of the international airlines went out of their way to generate media or rather aviation reporter loyalty.

Instead of the boring press release to launch a new product British Airways decided one day to fly us all down to Cape Town for Lunch.

When we arrived on their B747-200 at the airport in the city, we were immediately driven to the restaurant, had lunch and then back to the airport to board the flight back to Johannesburg.

While we were wined and dined the B747-200 was refuelled before the same plane departed for London via Johannesburg.

The irony about this press launch was we spent more time in the air than at the restaurant. Flying time between Johannesburg and Cape Town was about 2 and a half hours each way.

There were probably other PR stunts, but I can't remember any others.

Some of the PR departments had no idea what newspapers needed. While some airlines thought there announcements were earth-shattering, but many were mundane.

Chapter 16

Many people think the life of an aviation reporter is just fun and games. Certainly even in the newsroom, some colleagues thought this was the case.

Yes, you might get to fly to overseas destinations, yes you will travel business class, but believe me, most airlines will get their pound of flesh out of you.

The first overseas trip as an aviation reporter came about thanks to South African Airways (SAA) and Boeing.

The airline had placed an order for several new B747-400, and 17 aviation reporters were invited to travel to the Boeing factory near Seattle.

We were going there to pick up the new plane, and spend 10 days learning all about Boeing, but best of all visit their Everett factory.

We first flew to London with SAA on a B747-300, and after an overnight stay in London we left the next day for Seattle on a British Airways B747-200.

Both flights were business class. This was my first time to turn left on entering an aircraft.

What a difference turning left makes.

The food was better, the chairs bigger and wider, and unlike in economy class or at the back of the aircraft, passengers were not handed out trays of pre-cooked food. Another difference was Champaign was available to those passengers including us.

For a boy from Africa the opportunity to drink as much French Champaign as he wanted life couldn't get any better.

It goes without saying on both flights vast amounts of Champaign were enjoyed.

One of the journalists decided he was going to pack 6 bottles of good old South African brandy in his luggage. He wasn't going to drink the rubbish served up at the hotel.

Once we landed in Seattle, we were taken to the hotel, and for the next few days it was time for work.

In between various lectures and visiting the Boeing factory, we were given one afternoon off to do some shopping.

Boeing made sure this trip was not going to be a free holiday.

The lectures included facts about the new B747-400 as well as 2 lectures on Boeing's future into supersonic aircraft. One of the other lectures dealt with the history of Boeing.

The highlight came when we were all taken to the Boeing factory in Everett on the outskirts of Seattle.

At the time this factory was the largest building by volume in the world covering 98 acres.

Unlike visitors to the factory we were allowed on to the production and   assembly floor.

The 90 minute visitor tour only allowed people to watch from balconies above the assembly line.

After their trip these visitors were able to get something to drink or eat at the appropriately named paper plane café.

In January 1991 the factory assembled the B747-400 and the B767.

The plant employed more than 30000 people and the designers of the building worked out that the body heat from those working inside the building made sure no heating was needed even if the snow outside lay metres deep.

The massive complex spans both sides of state Route 526 named the Boeing Freeway but more about this later.

Travelling by sound and smell

Since that trip, the factory began production on the B777 and the B787 Dreamliner.

However, what with the pandemic and various other manufacturing issues, much production now takes place at their plant in South Carolina.

On 6 December 2022 the last of the Boeing 747 aircraft was rolled out.

Over more than 50 years later and 1574 747 aircraft were assembled at this plant. The last 747 was a 747-8 freighter.

The factory still is responsible for the 777 and some 787 aircraft.

Since its opening in 1967, more than 150000 people have visited the factory each year.

VIP visitors have included U.S. presidents, international dignitaries, CEO's, astronauts and other celebrities.

Before I stop talking about the plant, I have to mention the following interesting facts.

The length of the 747 is 225 feet long with its tail as tall as a six-story building and a wingspan that could accommodate 45 cars.

The Boeing Everett campus is big enough to encompass Disneyland with 12 acres left over for parking.

More than 30,000 people work at Boeing Everett, which has its own fire department, security team, day care and fitness centres.

It also has its own medical services on station, and a selection of cafes and restaurants to feed the thousands of workers.

There are also over 1300 bicycles to hire to get around the building quicker.

Overhead are a multitude of cranes used to move some of the heavier aircraft parts as the planes start to take shape.

There are six doors on the south side of the factory. The four to the west are 82 feet high and 300 feet wide. The two to the east are 82 feet high and 350 feet wide, and all can be opened with the

John CT Miller

simple push of a button and take about five minutes to completely open.

The murals on the factory's six massive doors are the biggest digital graphics on the planet, covering more than 100,000 square feet.

If these facts have not bored you, here are a few more about this incredible building.

The ceiling is 90 feet high above the factory floor, and enough to fit an eight-story office building inside. It has approximately 1 million overhead lights.

The plant has 26 overhead cranes that run on 39 miles of ceiling tracks, which lift and move big pieces and sections of planes as they're being built. The aircraft are assembled on a production line that moves about an inch-and-a-half per minute.

One thing the Everett factory doesn't have, is air conditioning. If it starts to get too warm inside, workers open the factory doors and use fans to draw air inside to cool the facility.

When it gets too chilly, they turn on more of the overhead lights to heat the inside air.

After the aircraft are fitted with their engines, they are towed over a bridge to the nearby runway.

One other interesting fact we were told was to get the planes across the highway the stretch of road is closed to vehicles for about 2 hours at midnight.

Obviously the planners did not think of that when they purchased the land and built the plant.

While visiting the factory our hands on tour was suddenly interrupted when someone came in and told us the Gulf war had just begun.

We all couldn't wait to get back to the hotel to turn on the TV and see what was happening.

Travelling by sound and smell

At the time in South Africa, the only international news station was CNN. In America there were several major news stations, each broadcasting live from Iraq and Kuwait.

All of us were glued to the various live TV channels apart from the journalist from SABC, the South African state broadcaster, who decided to go shopping instead.

A couple of days later we were ready to board SAA's first B747-400 named Durban as it made its way back to Johannesburg.

There were only about 60 passengers on board. We landed at Ilha da Sal to refuel and change the aircraft registration.

While on the runway, someone climbed up a ladder and ripped away the temporary US registration to reveal the South African registration.

About 17 or 18 hours after leaving Everett we arrived over Johannesburg, and before we could land, the plane did a circuit over the city.

It was good to be back home.

Chapter 17

My next overseas flight was thanks to Lufthansa in 1992 when a group of South African aviation journalists were invited to attend the opening of the new Munich airport, the first new airport in Europe in the past 40 years.

Before we left for Germany, the PR Karin in South Africa asked us if there was anything else we wanted to do in Germany.

I told her I would love to visit Berlin. A couple of years earlier in 1989 west and east Berlin was unified after the Berlin wall came down.

While open skies across most nations became the norm, cities divided by walls also became a thing of the past.

However, in the years that followed, gated communities and walls and fences between countries would appear.

On 9 November a spokesman for the communist part in East Berlin told its citizens they were free to cross into the western part of the divided city.

Thousands of people on both sides flocked to the wall and about 2 million people from the east side of the city streamed into west berlin.

People used hammers and any other implements to chip away at the wall before cranes and bulldozers pulled down section after section.

After the visit to Munich, my plan was to fly into formerly West Berlin, and fly out from the airport in the east part of Berlin.

Our group of South African journalists flew into Frankfurt on a

Travelling by sound and smell

B747-400, then transferred to a b737 and landed at the old airport in Munich.

The next day, we were taken on a guided tour of the new airport.

I didn't realise journalists from around the world had also been invited to attend the opening of Europe's newest airport.

The terminal and its facilities were pretty impressive. But I guess this is what one expects from German efficiency.

That same day we went back to the old airport to watch the last commercial flight take off.

The following day we went back to the new airport and along with various dignitaries witnessed the official opening of the airport.

Part of the opening ceremony included a take-off of the new Airbus A340 along with the prototype of the Do 238 Dornier.

I remember at the time thinking how quiet the Airbus was.

Talking about Dornier, we also visited their plant at Oberpfaffenhofen on the outskirts of the city. The Do 238 was the first 30-seater regional jet and they were busy with airworthiness certification trials with the prototype at the time.

In a nearby park 3 aircraft were on display. These included the super constellation, the Douglas DC-3 and the Junker JU 52.

Within a few years the new airport terminal had reached its full capacity of over 20 million passengers and a second terminal was built.

The new airport included even a special terminal for el al offering maximum security to those passengers flying to Israel.

While we were in Munich, we visited the highly automated BMW plant.

It was amazing to see robots carrying bits of cars and making their way across the factory floor.

Who needs manpower when these more efficient subjects can

work 24 hours a day and never complain, strike, ask for time off or demand a pay rise?

It goes without saying part of our visit included eating and drinking at one of the traditional beer halls.

Another night we were taken to a vodka bar. Here we met up with some journalists from Poland.

These journalist thought they were going to teach us South Africans how to drink Vodka.

They were proved wrong, and I was so to speak at the end of that night, the last man standing.

Karin our Lufthansa PR went out of her way to accommodate us all during that trip.

We also visited a centre where we got to sample different beers from various parts of the country as well as regional foods.

Following our visit to Munich, My next stop was Berlin.

Karin had managed to organise a guide to accompany me during my few days in Berlin.

I deliberately flew into Tempelhof in the western part of the city, and departed from Schonefeld, the only airport in the former east Germany.

Chapter 18

Just like many shops and buildings the difference between the two Berlin terminals west and east were stark and still noticeable some 2 years later even after the wall came down.

I naturally had to go to the famous Brandenburg gate or what was left of it, and touch the remains of the wall that divided people for so many years, during which so many people died trying to escape to the west.

My guide told me something that has stuck with me for many years. He said only the best cement was ever used on the wall, while houses and other building got second grade cement.

The Berlin Wall was built in 1961 to stop an exodus from the eastern, communist part of divided Germany to the more prosperous west.

I can never understand why communist governments around the world which claim the benefits of communism most times refuse to allow their citizens to visit the terrible west?

Ironically, these days some 30 years later, there are countries building walls and fences to stop others from coming into their country.

Most communist ruled countries still make it extremely difficult for their citizens to leave their country let alone holiday outside the country.

Between 1949 and 1961 more than 2 million East Germans out of a total population of 17 million, managed to escape to the West. Many were skilled professionals and their loss was increasingly felt in the German Democratic Republic, or GDR, as it was called.

John CT Miller

With the country on the edge of economic and social collapse, the East German government made the decision to close the entire border, and on August 13 1961 overnight erected the wall.

This date was deliberately chosen because it was a Sunday during the summer holidays.

It was often referred to by eastern authorities as the anti-fascist protection barrier, to protect East Germans from the west.

The concrete wall complete with 300 guard towers at regular intervals, was 96 miles in length and 13 feet high, though to start with it comprised temporary barriers of barbed wire coils.

Over days and weeks the barbed wire was replaced with vertical concrete slabs reinforced with iron bars and hollow blocks.

Behind the wall the so-called "death strip" contained anti-vehicle trenches, beds of nails and other defence devices. Wherever the boundary ran through water, similar defence mechanisms were put in place to prevent anyone escaping.

At least 138 people lost their lives trying to escape across the wall, but an estimated 5,000 managed to flee.

Those who did managed to escape hid in cars, sneaked through border points, crashed tanks through the fortifications, swam across the Teltow canal, paddled on a lilo over the river Spree, or crawled out via tunnels specially constructed by teams of dedicated volunteers including would-be people wanting to escape.

Among the most spectacular was the circus tightrope walker who walked across a disused power line to the west, breaking both arms in the process.

Most of those who tried their luck were males with an average age of 25.

In the early evening of 9 November, a government spokesman told a press conference that East Germans would be free to travel into West Germany. Asked when, he hesitated, and to the shock

Travelling by sound and smell

and amazement of the Germans present, added: "immediately".

As soon as western media reported the border had opened, people started gathering in large numbers at checkpoints on both sides. Overwhelmed by the numbers, passport checks were forgotten about by guards at around 11.30pm, by which time people were surging through.

It wasn't until 11 and 12 November when the first pieces of wall were pulled down. A hole was made in the wall segment cutting off the Brandenburg Gate on 10 November, but then sealed off again by the East German authorities and the wall did not come down properly until 22 December.

In the days after 9 November, east German authorities initially began removing pieces of the wall using angle grinders, construction vehicles and cranes, to create more crossing points between east and west.

Thousands of people known as wall peckers with hammers and chisels also came to take pieces home.

Later, people would rent out the hammers for a fee. The thud and chink sound of the hammers on the iron-reinforced concrete was heard for months.

Most segments remained in place and it took more than two years to remove the vast majority, with the official demolition programme not starting until the summer of 1990.

After much debate about the best way to mark the position of the former wall, a double row of cobblestones were discreetly set into public streets and pavements.

Sections still remain but there is said to be more of the wall on display in the US than in Berlin itself.

Hundreds of wall segments have been shipped to more than 50 countries, mostly as commemoration pieces and acts of solidarity and friendship.

Sometimes auctioned pieces of the wall were put on display in private estates.

Pieces of the wall can now be found as far afield as the north-south Korean border, a train station in Monaco, a urinal in Las Vegas and a historic east-west summit venue in Reykjavik.

Segments of the wall found their way overseas in interesting ways. One ended up in Kingston, Jamaica, after being gifted to Usain Bolt following his record-breaking 100 metre dash in 2009.

I wonder how many people in Cape Town even know The only piece of the Wall on the African continent is in the mother city? It can be found at St. Georges Mall, a pedestrian area in the centre of the city.

When Nelson Mandela visited Germany in May 1996 on a state visit, he received this special gift from the city of Berlin.

A piece of the Berlin Wall has stood in front of Harmonie, a German club in Australia's capital since 3 October 1992. The club had decided to commemorate German reunification on its grounds.

Three sections of the Berlin Wall stand on Plaza de Berlin in Guatemala City as part of the "Berlin por la Libertad" monument. The three sections of the Wall stand on concrete pedestals in a pool of water with colourful reliefs.

As a reminder of the Fall of the Wall, the last of the three pieces is turned on its side.

Several pieces of the Berlin Wall can be found in Fatima, the most important Catholic pilgrimage site in Portugal. They came as pilgrims' gifts to this place; one piece of the wall has been kept inside the shrine since 1991.

In Honolulu Hawaii, a piece of the Berlin wall made its way to the campus of Hawaii Community College.

The students and professors contacted the Berlin Senate in 1991

Travelling by sound and smell

with a request for a piece of the Wall.

In Montreal, the second largest city in Canada, a colourfully painted section of the Berlin Wall is one of the attractions of the city's World Trade Centre. It was given as a gift from Berlin in 1992 to mark Montreal's 250th anniversary.

In Osaka's Tokoku-ji temple, two grey sections of the Berlin Wall are nestled among the curved roofs, bonsai trees and stone Buddha's. The temple maintains close relations with Korea and reunification is a central focus here.

There are so many pieces of the Berlin Wall in the United States.

One of the most beautiful places is probably the piece of the Wall in front of the Ronald Reagan Presidential Library, which stands on a hill high above the city.

In good weather, you can see from here to the Pacific. A butterfly has been painted on a piece of the Wall that once faced West Berlin.

Three sections of the Berlin Wall can also be found in Seoul. In 2005, the Wall pieces were given as gifts from the Berlin Senate to the Korean capital.

In 1990, then Pope John Paul II received a piece of the Berlin Wall from Italian businessman Marco Piccininni.

The piece is now located in the northerly section of the Vatican Gardens.

Even on Mars, there's a piece of the Berlin Wall, even if no human took it up there. When the red planet was being mapped, an 85 cm boulder was named "Broken Wall" in commemoration of 9 November 1989. Later, other rock formations received names including Monday Demonstration, Nicholaikirche and Reunification.

I just wonder in years to come, how many people will know about let alone learn about the Berlin wall?

Chapter 19

On my tour of east Berlin, we went into quite a few shops. They were all pretty small, drab and uninspiring.

It soon became apparent customer service had not reached that side of the city. The shop keepers staff were obviously not interested in customers or for that matter customer service.

While in Berlin, I also made a quick visit to the so-called forest a green belt on the western side of the wall.

When it was time to catch my flight out of Berlin, I wasn't at all surprised at the difference between the 2 airports.

Tempelhof in the west was modern, while Schonefeld in the east was drab, lacking air conditioning and rundown, and once again customer service had not reached that airport.

Chapter 20

My final free overseas trip was thanks to British Airways on Concorde to New York.

As a badly paid journalist, flying business class was never on the cards, let alone even thinking about ever flying on Concorde.

On my various trips leaving Heathrow, I often waited in a plane on the runway when I could hear and feel Concorde as it took off with its afterburners on.

The next time its afterburners would be used was breaking through the sound barriers.

Now it was my time to make other passengers wait when we took off.

Concorde the first supersonic civilian aircraft was a joint venture between the British and the French.

Let me give you some facts about this engineering marvel.

Before I do, remember a single-seater jet fighter was able to reach speeds of Mach 2: twice the speed of sound. But British and French engineers had somehow managed to build a passenger aircraft carrying 100 passengers on board travelling at Mach 2.04 1560 miles per hour.

This was and remains the case to this day: a truly incredible engineering feat.

Its speed meant Concorde's fuselage would increase by almost 9 inches due to the heat.

It also meant a trip between London and New York instead of about 8 hours would now take just over 3 hours flying time.

Another way of explaining this was by using the phrase at the time "Concorde passengers arrived before they left." In other words if you left London at 4pm, you arrived in New York at 2pm local time.

I wonder if the 2.5 million passengers lucky enough or in most cases rich enough to fly on Concorde between 1976 and 2003 ever thought they would be sitting in a plane flying at twice the speed of sound and 60000 feet above the surface of the earth and enjoying a meal?

I certainly didn't.

If passengers looked above them, the blackness of near space was visible while below them, they could see the outline of the curvature of the earth.

While airline meals first started back in 1919, none have matched Concorde's for total extravagance or achieved quite the same wow factor. Concorde offered the only opportunity to eat at the edge of space: About 60000 feet or about 11 miles above the earth.

We arrived in London having flown business class on board a British Airways B747-400.

Before our 6 journalist boarded Concorde, we attended a talk about the history of this defying gravity plane.

Not only did Concorde have right of way on the runway at Heathrow, but Concorde passengers enjoyed a separate and exclusive lounge area.

This exclusive lounge also offered a secretarial service, and the large windows looked out over Concorde parked just outside the lounge.

The first passenger flights took place on 21 January 1976. On that day, two planes took off simultaneously. A British Airways flight flew from London to Bahrain, and an Air France Concorde from Paris to Rio de Janeiro via Senegal.

Travelling by sound and smell

Once on board, I don't know what I expected, having now been used to the amount of space on jumbo aircraft.

Yes, all seats were leather, but the cabin size took me back to my Viscount days.

Apart from the leather seats, another difference compared to jumbo jets, there were no cabin movies or back-of-seat screens just a Mach display showing passengers the speed, outside temperature and altitude.

The menus which were changed every week always included canapés, at least three courses of sumptuous food, high-end cocktails and an impressive wine list. Vintage champagne for brunch and heaps of caviar was the norm.

Concorde gourmet meals were served by 2 galleys and it also had its own wine cellar with the best taken from French vineyards.

Most of the stock had been purchased years before that first commercial flight.

A red, white, champagne and port were selected for each flight. Passengers could savour vintage champagne, claret, whisky, cocktails and liqueurs: All included in the price of the ticket.

While British Airways menu designs were mainly chic but simple, Air France menus were often intricately illustrated with images of hot air balloons, French characters and stylised planes.

Although Air France and British Airways ran different operations, and the dishes on each airline's menus differed to some degree, the food was largely French cuisine, with some English classics such as game pie and a full English breakfast. Menus from both airlines were usually written in French and English.

On that first flight, Concorde passengers flying from Bahrain to London in January 1976 were treated to caviar and smoked salmon canapés, cold breast of chicken with foie gras, asparagus spears and oranges poached in Grand Marnier.

John CT Miller

Menus on the British Airways Concorde always included Dom Pérignon champagne, often with lobster and caviar canapés, fillet steak, palm heart salad with roquefort dressing and fresh strawberries with cream.

Towards the end of Concorde's reign in the sky, renowned Michelin-starred chefs all contributed to British Airways Concorde's in-flight menus.

On 22 November 1977, British Airways' inaugural flight from London Heathrow to New York's JFK airport was such an event that menus from the flight now fetch more than £1,000 ($1,400) on online auction sites.

On a 1982 British Airways London–Washington flight, which took a mere 4 hours and 5 minutes (a few minutes more than the flight to New York), the menu boasted champagne, a 1979 Chablis and a range of high-end liqueurs.

The starters included canapés of galantine of chicken, caviar and smoked salmon; and poached fresh crawfish garnished with a lobster claw.

The mains consisted of a choice of pan-fried veal steak, fillet of turbot with lobster sauce, or venison steaklets with three types of vegetables, including a salad. A selection of English and French cheese and a dessert of woodland berries and cream was next, then coffee, cakes and homemade chocolates concluded the meal.

On the return journey from Washington to London was more of the same. As well as plentiful drinks there were canapés of caviar, goose liver pate and shrimp, and smoked salmon with crab legs.

Next was a choice of prime fillet of beef with sautéed chantarelles, an English style game pie, or trout Cleopatra and vegetables. The menu also offered a selection of French and English cheeses, strawberries Romanoff and coffee with chocolate mint crisps.

Meanwhile on Air France flights, Concorde passengers also ate

Travelling by sound and smell

the best food money can buy – caviar and lobster salad with truffles then guinea fowl in champagne sauce, washed down with Cristal champagne.

A brunch menu from Concorde's final year included an appetiser of pineapple carpaccio with summer berries. Passengers could then choose from a full English breakfast, medallions of beef with black truffle sauce, an open lasagne or a poached salmon terrine. Dessert was a baked pear with prunes in Armagnac crème Anglaise, or cheese.

Bread rolls, pastries and coffee were also available. If the menu proved too much guests could have a modest sandwich, not that Concorde meals could ever be described as modest.

The choice of tableware was just as important.

Royal Doulton, Conran and Nachtmann were some of the companies that designed quality fine bone china, silver plated cutlery and crystalline glassware for British Airways Concorde flights. Air France treated its guests to tableware of a similarly high standard.

The 24th of October 2003 marked the end of an era when Concorde flew for the last time. Flight BA002 took passengers including actress Joan Collins and Sir David Frost from New York to London. Aviation aficionados can buy the classy silver menu from that final journey – it's currently available to buy on Ebay for around £2,000 ($2,700).

The last meal on Concorde included naturally Champagne.

Breakfast started with three vintage champagnes and Scottish smoked salmon with caviar. Then came options including a mixed grill of pancetta-wrapped prime fillet of beef, lamb cutlets, lobster fishcakes with bloody mary relish, a wild mushroom and truffle omelette and Greek yoghurt with fruit. A buttermilk pannacotta dessert and cheese finished the meal.

John CT Miller

Every passenger who flew on Concorde received a gift to mark the unique occasion of flying on a supersonic plane. These included such Souvenirs as Wedgwood paperweights and circular trays, Smythson of Bond Street notebooks, silver photo frames, letter openers, leather drinks coasters, hip flasks, leather bags, Concorde prints and flight certificates.

Before smoking was phased out on flights in the late 1980s and finally banned from Concorde in 1997, passengers were also offered Havana cigars.

I cannot remember the 2 meals we enjoyed, but I do remember at one stage touching the window, and was surprised to find it was quite hot. Certainly, before and since then, I have never found windows on any aircraft to be that hot.

Something else I remember about those 2 flights was not knowing or feeling when we went through Mach 2. If one of my colleagues had not told me, I would not have known. All I felt was a slight bump.

I doubt if passengers would have known if it had not been for a speed indicator at the front of the cabin displaying the velocity.

Out of London, even Mach 1 was not noticeable. However, leaving New York and immediately flying over the ocean it only took a few minutes before we broke the sound barrier.

Of course the other thing I could not help but notice was on take-off how passengers were thrust back into their seats.

We were all given a chance to spend a few minutes in the cockpit talking to the flight deck crew.

Who ever thought it would be less than 10 years before cockpits across the world on all aircraft would be locked following the New York terrorist attack in 2001.

I remember writing an article comparing Concorde's lifespan similar to the DC 3.

Travelling by sound and smell

Perhaps I should explain.

More than 70 years after the DC 3 took to the skies, I could see at the time the same thing happening with Concorde. Compared to the amount of hours each year jumbo jet engines were in action, Concorde's engines did not do a lot of work each year.

Little did I and other people ever dream it would also be about 10 years later, when Concorde would no longer take to and be seen in the skies.

There were 20 Concorde aircraft built. 14 aircraft were shared between British Airways and Air France.

On July 25 2000, 2 minutes after take-off an Air France Concorde crashed killing all 109 on board and 4 people on the ground.

During its take-off, the plane struck a piece of metal the size of a small penny on the runway, which punctured one of the tyres.

Pieces of the shredded tyre struck and ruptured one of the fuel tanks causing a massive fire.

In 2001 in November, the Concorde returned to the air, but 2 years later British Airways and Air France without warning announced Concorde would no longer fly.

BA refused to sell its Concorde aircraft to Virgin Atlantic its main opposition.

I still believe BA had no right in deciding this as the money to build the only supersonic passenger jet came from the tax payer and not the company.

Chapter 21

Our brief stay in New York included a few talks on investment and economic opportunities as well as visits to several landmarks including the World Trade Centre.

Little did we know at that time, within less than a decade the New York landscape would change for ever.

We were lucky enough to be treated to a meal at the famous Windows of the World restaurant on the 107th floor of the World Trade Centre, this once iconic building.

The main dining room faced north and east, allowing guests to look out onto the skyline of

Manhattan. The dress code required jackets for men and was strictly enforced. A man who arrived with a reservation but without a jacket was seated at the bar. The restaurant offered jackets that were loaned to the patrons so they could eat in the main dining room.

Wild Blue, a more intimate dining room was located on the south side of the restaurant. The bar extended along the south side of 1 World Trade Centre as well as the corner over part of the east side.

Looking out the full length windows from the bar, patrons could see views of the southern tip of Manhattan where the Hudson and East Rivers meet.

The kitchens, utility spaces, and conference centre in the restaurant were located on the 106th floor.

After we took our seats at the restaurant, I ordered a giant lobster. I decided this would probably be my only chance to eat a 2.5 kg lobster.

Travelling by sound and smell

When it arrived, I could not believe the size. It was enormous and took up the extra-large plate.

Little did I know that when I returned to New York some years later, the World Trade Centre buildings would no longer be there.

While we were in New York, another memorable visit was to the New York stock exchange. The timing was perfect. We heard the ringing of the bell from the balcony.

Since 1792 the New York stock exchange has only closed on a few occasions. One included when hurricane Sandy struck the city.

The stock exchange started on May 17, 1792, when 24 brokers and merchants met under a buttonwood tree.

It was under this tree that these two dozen individuals signed the Buttonwood Agreement, establishing the grounds for trading at what was then called the New York Stock & Exchange Board.

That name was shortened in 1863 to what we know today: The New York Stock Exchange.

To commemorate the document leading to its creation, the stock exchange planted a new buttonwood tree out front.

The stock exchange has seating for 1366 people, and if and when a seat is sold it goes for millions.

Although the trading floor isn't as lively as it once was — with the decrease of people running around and volume of open outcries Currently, there are approximately 500-1,000 people trading on the floor each day.

Unfortunately, the exchange is no longer accessible to the public.

After the September 11th attacks and the increase in security that followed, many buildings once open to the public were severely limited or shut down altogether. For the stock exchange, it meant no more public tours. A fence was even added to the front of the building to prevent pedestrians from getting too close.

Before 9/11, visitors could book tours of the building and see the trading room floor in action.

Nowadays, the only people allowed inside are brokers.

The closest visitors can now get is in front of the fence.

Despite the new security measures, there are still many ways to experience the New York Stock Exchange without going inside.

Every weekday to mark the beginning of trading, the opening bell of the NYSE is rung at 9:30 a.m. and to mark the end of trading the bell is rung at 4 pm.

When the guest ringing the bell fails to ring it for the acceptable amount of time — 10 seconds for the opening bell and 15 seconds for the closing bell – it's not uncommon for the floor to erupt into boos.

In the past, the bells used to be rung by floor managers, but they later started inviting executives, public figures, and celebrities to ring them, which became a daily event.

On special holidays, like Independence Day, the American flag hangs across the building.

Every year, there is a Christmas tree lighting on Broad Street, in front of the New York Stock Exchange. Lights are coiled around the exchange's Corinthian columns and around a massive pine tree.

Another place we visited was the Blue Note Jaz club.

Within a few years after the Blue Note Jaz club in Greenwich opened in 1981, it became the place to go to and be seen.

Some of the world's best known jazz artists have performed there.

Artists like Dizzy Gillespie, Sarah Vaughan, Canadian born drummer David Mendel, Carmen McRae, Dan Frieber, Lionel Hampton, Oscar Peterson and the Modern Jazz Quartet.

Another famous artist to have appeared there was Ray Charles,

Travelling by sound and smell

who did so for a week every year.

The club is one of the few which boasts its own record label called Half Note Records, and Blue Note entertainment has opened clubs in many cities across the world.

Someone who I interviewed in South Africa some years before our trip Chick Corea also became a regular performer at the club.

I am not a jazz lover, so the club for me was not my first choice. I would much rather have attended a performance on Broadway, but this was not to be.

I would have to wait a few years before attending a performance on Broadway.

While in New York we were also allowed a few hours shopping time before flying back to London on Concorde.

That afternoon I went to Bloomingdale's and Macy's. I can't remember buying anything, but do remember thinking how old the buildings were. There were even wooden escalators still working on certain floors in Macy's.

I think much like Harrods in London, you had to buy something if only to be seen with one of the quintessential and recognisable shopping bags.

For me, the flight out of New York was far more impressive when it came to feeling speed.

I seem to remember it was only a few minutes after we took off we reached Mach 1, and I actually felt the plane break through the sound barrier.

When we arrived back in London, I stopped over for a few days.

I had previously organised to interview Richard Branson. At the time there was talk of Virgin Atlantic beginning regular flights to South Africa.

It had taken months to set up the interview thanks to the then head of the communications department Will Whitehorn.

John CT Miller

It was during the interview Branson asked me if I had ever flown on Virgin Atlantic.

The next day I did the 16 hour round trip on a B747-200 from Gatwick to Boston and back.

On my trip back to south Africa I unfortunately had several to many vodka and tomato juice drinks, which did not impress the British Airways Johannesburg office. Forget about the series men behaving badly, I certainly did on that flight back.

Chapter 22

With Concorde and travelling at twice the speed of sound a thing of the past my next flying experience couldn't have been more different.

I was lucky enough to be invited along with a few other aviation journalists to be part of a rather unique flying experience.

A group of Swiss were in the country to fly on 3 different aircraft each over 60 years old in a single day.

This novelty trip was more about bragging rights for the Swiss.

Each flight lasted about 30 minutes.

The morning started when we boarded the DC 3 and flew over Johannesburg. When we landed, we boarded a DC 4 and after that took to the air in the Junkers JU 52.

My memory of this last plane was sitting at the back of the aircraft and able to open the small window.

My next inflight experience was when I was invited out by a member of the South African aerobatics team one Sunday morning.

The pilot had arranged to pick me up at the Star newspaper building.

When he arrived and we met, he was somewhat taken aback to discover I was blind.

After that hurdle was overcome, he happened to mention he hoped I had not partied too much the night before.

He was wrong. I was suffering from a hangover. Naturally I didn't let him know this, but I did wonder how my body would react to flying upside down and doing various loops.

We arrived at the airfield and after been strapped in to the harness, we took off.

I guess for a sighted person, such an experience would be either inspiring or terrifying.

All I can tell you for me it was neither. The best way to describe my experience is to say it was interesting to fly upside down and experience the g-force during some of the manoeuvres.

I can report back that fortunately I did not throw up during or after that flight.

You've all heard off friggatriskaidekaphobiaor the long-held superstition fear of Friday 13$^{t?}$

Well in attempt to find any excuse to fly, I convinced the news editor I would write a story about boarding a plane on Friday the 13th at 13 hundred hours.

He thought this might make an interesting story for the following day.

Remember these were the days before the terrorist attack on the twin towers, and sometimes passengers were lucky enough to be invited into the cockpit.

We took off in the A320 at exactly 13 hundred hours for Durban, and after a 45 minute stopover we flew back to Johannesburg with me sitting in the cockpit.

The pilot asked me if I wanted to "watch" the plane land on auto-pilot?

Everything went well until final approach and we were within minutes of landing, when a gust of wind blew the plane off course, and the pilot had to take control and we did another circuit before landing.

I remember thinking at the time, if only the passengers had known this delay was all my fault, I don't think they would have

Travelling by sound and smell

been too happy.

A few years later, I would share an on-ground experience but not as a passenger but with probably millions of people around the world.

This experience was shared with over 60000 rugby fans on June 25 1995.

It was the Rugby World Cup final between the South African national team the Springboks and the New Zealand team the all blacks at Ellis Park in Johannesburg.

At just after 2.30pm to the exact second the spectators at Ellis Park in Johannesburg couldn't believe their ears and eyes as a SAA B747-244 flying at about 300 meters flew twice over the stadium.

Before that final, the captain of the plane Laurie Kay and his crew had spent hours in a flight simulator. They trained overflying their imaginary route again and again to be sure of crossing the stadium at the very second they were expected.

The simulator software did not include computer-generated graphics of Ellis Park and instead a graphic of an airport terminal was used.

SAA needed special permission from the civil aviation authority that day.

Most people don't realise the jumbo was crawling along at a mere 230km/h however, a jumbo can slow down to less than 180km/h before dropping out of the sky.

Captain Kay later said the plane wasn't as close to the ground as it looked: A jumbo covers over an acre of ground, if measured from wing-tip to wing-tip and nose to tail.

Overflying the stadium wasn't an issue, but because of the live TV broadcast, the flypast had to be at the exact second.

Kay said he and his team agreed with the broadcaster they would fly over at exactly 2.31 and 45 seconds.

John CT Miller

When his team asked how much leeway they had, they were told there was no leeway.

The broadcaster also asked for two passes, and Kay and his team were told they had 90 seconds between them.

Apart from those involved in the aviation world, the public had not heard of captain Laurie Kay, but after that most South Africans, and I have no doubt many around the world knew about Laurie Kay.

I later had the privilege and I mean the privilege of interviewing this remarkable yet humble pilot.

He was not just passionate about flying, but also teaching flying.

He told me his love of all flying things began when he was about 6 years old when he joined the South African Air Boys Club in Malvern Johannesburg. He often hitched a ride on an older boy's bicycle to the club.

As he got older the passion didn't dwindle, if anything, it became stronger and by the time he left school all he wanted to do was fly.

While listening to him, I wondered if it had not been for the accident, would I have become a second Laurie Kay?

Well, maybe not a second Laurie Kay, but certainly a pilot who lived for flying.

Laurie told me he originally applied to join the South African Air Force but was rejected.

Finally, the Royal Air Force accepted him and he commenced his pilot training in 1967 at RAF Church Fenton, on De Havilland Chipmunks.

On his return to South Africa Laurie joined the SAAF on a short service contract and was stationed at 5 squadron in Durban, where he did a conversion on his beloved Harvard. His love of this plane lasted his entire life until he sadly died many years later.

It wasn't long before he took up a post as an instructor at Central

Travelling by sound and smell

Flying School Dunnottar. While at Dunnottar Laurie instructed many of the people that would shape the future of aviation in South Africa.

Laurie made the big switch from military to civil aviation in 1974 when he joined South African Airways as a third pilot on the B707.

Twenty months later he moved to co-pilot on the B727 fleet and after that to a co-pilot position on the B747.

Laurie later returned to instructing, and became a training Captain on the Hawker Siddeley 748, B737-200 and the Airbus A330.

His love of all things Boeing continued especially with the B747 aircraft when he became a senior training captain on the 747SP and the classic 200 and 300 variants.

He told me once in the cockpit he was most content and at peace with the world.

I would be lying if I didn't admit, I was more than envious. I might never have been as great as Laurie, but I understood his love of flying.

This aviator will always be connected with major events in South Africa. Apart from the rugby World Cup, he also headed a 3 B747 plane formation fly past when Nelson Mandela was inaugurated as president of South Africa.

For him, it wasn't just civilian aircraft, aerobatics also played a very big part in Laurie's flying career.

In 1996 Laurie flew around the country in the brightly painted Olympic Boeing 747-300 which took the South African team to Atlanta.

This incredible man had 3 loves in his life. Flying, his wife and family and his dogs.

Laurie retired from commercial flying with well over 23000

John CT Miller

hours in his logbook.

He never stopped flying, and was active at the Harvard club of South Africa taking people for introductory flights and doing the odd Harvard display.

After I left South Africa to settle in the UK I learnt one of Laurie's projects was the distribution of a DVD "Flight for a Nation" covering all the major 747 displays he took part in.

The proceeds from the DVD all go to the SA Guide Dog Association and is available at 011-705 3512/3 at R120-00.

In April 2013 he died aged 68 at an air field in the Kruger Game Reserve.

At the time of his death, he had gone back to the game park to help with a rhino anti-poaching flying programme.

Chapter 23

In 1997, I had no idea that my next flight was going to be one of my last 3 trips with SAA.

The airline had earlier formed an alliance with Uganda Airways and Air Tanzania.

On August 30 I took off from Johannesburg in a B747 SP on my way to London via Uganda.

One of the amusing things about this aircraft it had an image of a male lion's head on the tail fin.

Now I know the lion is considered to be king of the jungle, but king of the skies? This was something else.

We arrived at Heathrow in London in the early morning of 31 August.

I will never forget that day. When we landed and while waiting to clear customs, one of the ground crew told me that princess Diana had died that morning.

To say I was shocked, would be an understatement.

As they say, the rest is history and it would be some 3 years before I returned for a brief visit to South Africa.

Chapter 24

In December 2000 I returned for a hasty visit to South Africa after learning my best friend for 30 years Kevin Mason was seriously ill with cancer.

The best deal at the time was with Swiss Air. However, it meant departing Heathrow at 6am.

For most with their own transport this would not have been a problem, but thanks to London Transport, there were no trains after midnight until 5am from central London to Heathrow.

The last train to arrive at Heathrow was at midnight.

I got the last train and spent the next few hours alone in a deserted airport.

At 6am I boarded the Swiss Air B737 to Zurich for the connecting flight on a B747 SP to Johannesburg.

This would be one of the last flights for that plane before it was scrapped.

It was also my only long distance daytime flight to South Africa.

Once we landed in Johannesburg, I boarded a B737 flight to Durban with Kulula one of the low cost airlines in South Africa.

This was going to be my final visit to my dear friend Kevin Mason a friendship lasting over 30 years.

The last time I had "seen"" Kevin, was when we went to see the Rolling Stones play at Ellis Park, the rugby ground made famous after the fly-past by captain Laurie Kay.

Some months before my flight, Kevin told me he had cancer.

The following year Kevin died.

Travelling by sound and smell

When I arrived back the editor of a website dedicated to Rhodesian musicians contacted me and wanted to interview me.

Unbeknown to me, Kevin had spoken to the editor Pete Shout about me.

Here is the article which appeared on the website.

"John Miller Musician and journalist.
Present location: near Cardiff in Wales.
John was born in Salisbury in January 1951 and was blinded in a shotgun accident as a 10 year old. This traumatic event drastically changed his life and he went on to attend King George VI primary school in Bulawayo for 3 years before continuing his secondary education at the School for the Blind in Worcester, South Africa. He matriculated in 1969.

John's interest in music took hold shortly after he lost his sight.

His first experience of a live gig came in 1961 when his grandmother, who was far more "cool" and adventurous than his disapproving mother!, accepted an invitation from Mickie Most to attend a couple of sessions in Salisbury.

Like so many aspiring musicians of the era, John fell under the spell of the Shadows and later, thanks to the kindness of the Cyclones' lead vocalist Dyllis Stevenson, he met Cliff Richard and the Shadows, as well as Carole Gray, the Bulawayo actress who starred opposite Cliff in the Young Ones.

A couple of the Dyllis and Cyclone tracks are to be found on youtube.

These include:
Hard times ahead – *Dyllis Stevenson & The Cyclones*
http://youtube/c3bO44Fi09Y
That's what I thought – *Dyllis Stevenson & The Cyclones*

John CT Miller

http://youtube/iCrLimAjSYY
Ghost Riders in the Sky by The Cyclones
http://youtube/pH8TQZl-KaA

John's path of introductions to the rich and famous (probably more famous than rich, truth be known!) continued when he met Jeremy Taylor of "Ag

Pleez Daddy" fame. Jeremy had occasion to visit the Miller home and sat for a while with John who was learning his way about the guitar. John was inspired by Jeremy's enthusiastic support and encouragement, spurring him on to persevere with his guitar playing ambitions.

After working as a personnel consultant in Salisbury for a couple of years, John returned to South Africa. Relocating to Pietermaritzburg in Natal via Johannesburg he joined a band formed by old school mates in 1974. John made his debut with As You Like It on rhythm guitar. The leader of the band was a Northern Rhodesian, Ian Farrington.

About a year later John formed his own semi-professional outfit and called it Choice. This quartet played numerous gigs throughout Natal for the next 4 years, a highlight of which was playing at a Rhodie bash at the Durban City

Hall and backing Graham Boyle, formerly lead singer of the Gentle People, which had a number 1 hit with (Rain Rain Rain).

John then left the music scene and, in 1980, was offered a job as a journalist on the Johannesburg based Rand Daily Mail. Whilst working as a court reporter he managed to ease himself back into the music scene. With the assistance of the editor of Express Beat, Suzanne Brenner, he began freelance entertainment writing for the Express, as well as the Mail on occasion. The Express Beat was part of the Sunday Express newspaper.

In his role of music journalist, John was able to attend many

Travelling by sound and smell

concerts and shows with his press pass and he fully exploited these opportunities. During this period he interviewed virtually all South Africa's leading artists, entertainers, record producers and executives, including Clout, Ballyhoo,

Juluka, Patric van Blerk, Hilton Rosenthal and the late Emil Zoghby. He was a frequent visitor to local recording studios although, for most of the time, he was supposed to be writing up the details of some gory court case or another.

One of his biggest scoops was breaking the news that Queen would be playing at Sun City, as well as being the first to report the break up of Juluka. The latter article appeared on the front page of the last ever edition of the Rand Daily Mail newspaper published on 30 April, 1985.

Shortly before the closure of the newspaper, John also financed and produced a concept rugby album with English lyrics on one side and Afrikaans on the other. For those interested, there are 3 tracks from the album on youtube.

There Once Was A Sailor & Sing Us Another One
http://youtu.be/PIBUJuABxCs
Nou Nou Nou – Rugby Sing-A-Long
http://youtu.be/N6WgpxUIopA
Schoeman Was his Name – Rugby Sing-A-Long
http://youtu.be/8CsBM9IxbkI

In the early 80's he teamed up with Suzanne Brenner and they wrote and published the first South African Country & Western booklet called "Surprisingly Enough" (SA Country Music). John attributes most of his success as an entertainment writer to Suzanne Brenner.

In 1984 John was offered the position as Southern African correspondent with Billboard Magazine. Again, Suzanne's hand had played a role in this offer being made. He took up the post and filled it for 7 years until a shift in editorial policy unsettled him and decided to leave the position. It was at this stage that John decided to leave South Africa and he settled in England in 1997. During the interceding years he worked as a consumer journalist on the Star newspaper.

John continues to have deep links with the continent of his birth and has visited South Africa, the last occasion being shortly before the death of his close friend and renowned musician, Kevin Mason, formerly of Gentle People, Gate, and Lincoln with these 3 groups comprising many

Rhodesians both Northern and Southern. Since moving to the UK John has become a prolific writer of books, all of which appear on Amazon under John C T Miller. His latest books focus on travel and conservation both of which he is passionate about.

He says "unfortunately there is a similarity between the many

endangered species in South Africa and old Rhodesians, both of which are slowly dying out with only pictures and sounds left to remind those left behind how magnificent they were."

Chapter 25

In 2001 on September 9 just before 9am New York time I became part of a global audience as thousands around the world watched in horror as the situation in that city unfolded.

The first of 2 planes deliberately crashed into the north and south World Trade Centre.

At almost the same moment, I had just walked downstairs to make a sandwich. Walking past the living room to the kitchen, I happened to hear on CNN a plane had crashed into the Twin Towers.

I stood there and listened. I remember thinking and also seem to recall CNN at that moment also thinking it might have been a small plane.

A few minutes later a second plane crashed into the buildings, and all hell broke loose: America was under a terrorist attack.

Within the next one hour another 2 planes had been hijacked by terrorists forcing their way into the cockpits and overpowering the pilots and taking command of the aircraft before deliberately crashing them.

The air space above America was placed in lockdown for the next 2 days.

I remember standing and listening and thinking I could have been one of the thousands who died that day.

Some years before 6 journalists including me had visited the World Trade Centre.

I spent the next 3 days in front of the TV listening to all the reports.

Within days if not hours, all cockpit doors on aircraft across the world were locked.

The days of passengers wanting to visit the flight deck were now a thing of the past.

At the same time, airport security around the world was dramatically increased.

At major airports, any liquids or jell over 100 ml was confiscated, and other personal items carefully scrutinised and checked. Shoes and belts had to be removed and pockets emptied when walking through the various security areas.

These added security measures meant passengers now had to be at the airport 3 hours before the flight departed.

At the windows of the World restaurant all of the staff members working that day died.

While firefighters waited on the ground for orders a man nearby an overhang saw a body fall onto it.

When the second plane crashed into the buildings the plane exploded, followed by a fireball and debris as big as cars were sent flying in all directions.

The fires were reportedly burning at more than 1,000 degrees, and the air was so saturated with suffocating, toxic smoke that it was unbreathable. Within two or three minutes of impact, people were seen falling from the building.

Crowds on the ground saw people waving large sheets what must have been the restaurant's tablecloths, before more people began to fall from the sky: some holding hands.

Chapter 26

A few years later my former partner Helen and I began a series of flights to Europe, New York and South Africa.

The first was a flight on a BA B757 to Krakow in Poland.

This trip was a double winner.

To begin with, it was my first flight on the B757 and it was going to be my first trip to Auschwitz.

I had always wanted to visit Auschwitz the largest of the German Nazi concentration camps and extermination centres. Over one million men women and children were deliberately murdered and killed at Auschwitz.

You can read all about this camp on the internet, but I will say I was very disappointed. Immediately outside the camp venders had set up selling not just food, but many cheap novelty items.

We wanted to get a train from the town centre to the camp, but this was not possible and all visitors were forced to take a bus to the camp.

While I was glad to finally visit Krakow, my lasting memory was most shopkeepers had obviously not heard of customer service.

Another place we visited while in Krakowwas the famous salt mine.

On the days we visited Auschwitz and the mine, Poland was experiencing a heatwave: it was about 40c.

On entering the mine, it was a relief to escape the heat and enjoy the underground coolness.

The Wieliczka Salt Mine first opened in the 13th century and is now an official Polish Historic Monument and a UNESCO World Heritage Site.

The mine reaches a depth of 327 meters, and extends via horizontal passages and chambers for over 287 kilometres.

As we made our way gradually further down into the mine, I was completely surprised to find out there was an underground lake 135 metres below the surface.

Our guide told us in the past even pleasure boats were found at the lake.

The 3.5-kilometer visitors' route is less than 2 percent of the mine passages' total length and includes statues carved from the rock salt at various times.

Inside the mine, there was a chapel, and a reception room used for private functions, including weddings.

Another chamber had walls carved by miners to resemble wood, as in wooden churches built in early centuries. A wooden staircase provides access to the mine's 64-metre level.

As we made our way down the mine, at one stage I licked my finger and touched one of the walls just to make sure it was salt.

I also said to Helen, I was surprised those working in the mine had so much time to carve out the various statues. I wondered if this was all done in their spare time?

When we finally reached the lowest point, thank God there was an elevator to return us to the surface.

I don't know what we expected, but the rock salt is naturally of varying shades of grey, resembling unpolished granite rather than the white crystalline substance we find in supermarkets.

Our guide also told us that during World War II, the mine was used by the occupying Germans as an underground facility for war-related manufacturing. The Nazis transported several thousand Jews from the forced labour camps in Plaszow and Mielec to the

Travelling by sound and smell

mine to work in the underground armament factory set up by the Germans in March and April 1944.

However, manufacturing never began as the Soviet offensive was nearing. Some of the machines and equipment were dismantled, including an electrical hoisting machine from the Regis Shaft, and transported to Liebenau in the Sudetes mountains.

He also told us about some of the notable visitors which include Nicolaus Copernicus, Fryderyk Chopin, and Pope John Paul II.

Our next flight also with British Airways was on a B737 to Cologne in Germany to visit the annual Christmas markets.

After landing and booking into our hotel, we took the subway train to the centre of Cologne.

Leaving the station, we found ourselves almost opposite the city's cathedral lit up by a giant Christmas tree, surrounded by about 150 Christmas stalls.

We would soon discover in many ways this was not the best market compared to the many other Christmas markets.

It seemed if around every corner, there was another Christmas market.

I think there are about 6 or 7 different Christmas markets in and around cologne, with all selling various wonderful snacks as well as the traditional glühweinat each of them.

While walking through a couple of the markets, we couldn't help but try some of the various snacks on offer. Some of the best snacks were little egg dumplings tossed in cheese, potato pancakes, much better than any hash brown, and then there were the cones filled with fries and topped with mayonnaise.

Another Moorish snack were the little pizzas, a thin flatbread covered in white cheese, onions and bacon.

There were also savoury crepes and others with apple sauce.

One of my other favourites were the different German sausages including the famous currywurst.

Apart from traditional Glühwein other venders used cherry or raspberry to flavour this winter warmer.

There was also coffee, hot chocolate and even craft beer.

Another market we visited was the Angels market: enchanting and not to be missed.

There were clusters of beautiful stars suspended above in the trees, and performers dressed like angels.

The wooden stalls at this market were painted white with simple overhead decorations made of natural materials like twigs.

At this market you can even buy waffles shaped like the Cologne Cathedral.

Not far from this market was the Nicholas' Village Christmas Market with its wooden stalls which actually had its own chapel decorated inside with frescos. Nearby was a small house where a fun program for kids run by local artists and performers appeared to be on offer.

Some of the markets sold various hand-made wooden decorations and gifts. What a pleasure to get away from the cheap Chinese plastic decorations.

After visiting these markets we made our way to the harbour on the Rhine to have a look at a Christmas market on board a boat.

This was really disappointing. Once we boarded the boat, the venders had a captive audience, and it seemed each tried to get us to buy something from them.

We didn't stay long, and were glad to escape back to dry land.

My lasting memory of Cologne was not just the markets, but how clean the subway trains and the stations were. But that is German proficiency for you.

After Germany, we went on a shopping trip to New York.

The 4 day breakaway to the Big Apple was on a Delta Airways

Travelling by sound and smell

B767.

There were 2 things I remember about this flight.

The first was being served ice-cream half way during the flight, and also the middle aged cabin crew compared to some other airlines which employed much younger people and not forgetting the extremely often hostile and unwelcoming immigration airport officials when we landed in New York.

Talking about unfriendly and unhelpful staff, this was the same with those working at the various subway stations.

Something I had forgotten about New York, was steam rising up from parts of the city's sidewalks.

While in the city, we had to visit some of the world famous department stores like Bloomingdale's, Macy's and Saks Fifth Avenue.

Many believe New Yorkers are not very friendly, but this is certainly not true of shop assistants.

Unlike those in the UK, New York shop assistants always went out of their way to help.

Customer service was important and as the saying goes, the customer is king.

It is a pity those working at most American airports and subway stations do not follow their example.

Another thing sadly lacking at most New York subway stations is elevators unlike in London where almost all stations have elevators.

I don't know how those in wheelchairs cope?

Even in 2022 most don't have elevators, and it will take another 30 years before most are equipped with elevators.

For those who have read some of my previous travel books, will know I refuse to eat or drink at any American fast-food chain. I believe in always supporting local businesses.

This decision was to say the least impossible in America for obvious reasons.

However while we were in the Big Apple, I broke this rule by visiting McDonalds. I wanted to find out if the food on offer was the same as that served in the UK.

No it wasn't. The American chain offered a much greater variety.

One other thing we both could not believe was the huge portions of food at restaurants.

Talking about bigger and better in America, we had to visit the world famous restaurant at 854 7th avenue known as Carnegie Deli to sample their massive smoked corned beef and pastrami on rye sandwiches and their equally huge cheese cakes.

This iconic restaurant just off Broadway opened its doors in 1937, but sadly, it closed its doors in 2016.

Not knowing how big each dish really was, we each decided to order 2 pastrami on rye and 2 cheese cakes.

The restaurant's motto says "If you can finish your meal, we've done something wrong."

Well, they were right, and definitely did nothing wrong.

When our orders arrived, we could not believe the size of the sandwiches and the cheese cake.

Another New York highlight for us was attending a show on Broadway.

Here again just like shop assistants, the audience reactions were completely different to those in the West End. As to be expected, the Brits were more reserved.

While in the Big Apple, we decided to get the train to New Jersey to visit a shopping mall and the famous JCPenny department store.

I must admit, I was disappointed in the store. If you were looking for fashion, this was not the go to outlet.

Somewhere else we wanted to visit was Ground Zero. Sadly the construction of the National September 11 memorial at Ground Zero had not begun when we were there.

Travelling by sound and smell

We decided on our last full day during our visit to the city that never sleeps, to visit the statue of liberty and Ellis Island. Sadly on the day, it was raining and not the weather to be crossing over the water from Manhattan by ferry.

Oh well, it was a great visit, and good to be back in New York again.

Perhaps the next trip there would be more time?

A couple of years later Helen decided to surprise me with a birthday present: a trip to South Africa.

Talking about birth days and dates. I later learnt a certain young girl called Greta Thunberg was also born on 3 January.

The trip turned out to be the first of 5 visits to that country.

It was also Helen's first trip to the country, and only a few weeks before we left she had managed to pass her driver's licence.

Our first 2 flights were with South African Airways (SAA).

The first was on the A340-600.

The cabin service on that flight was great, but the plane was rather drab according to Helen.

When we landed in Johannesburg on route to Cape Town, the captain made the following announcement.

He said "welcome to South Africa, this was the safest part of your trip, please be careful when you leave the aircraft and travel in our country."

I am sure if Helen could have got the next plane back to the UK she would have done so.

Unfortunately, I had always spoken about the crime, the hijackings and general lawlessness in the country. Before we even left, she was already hesitant about visiting South Africa.

Our next trip was in business class on the (SAA) A340-300.

This was the first time Helen had turned left into business class on entering an aircraft.

Helen was most impressed with the food and facilities in the business class lounge at Heathrow, but we both found the lounge at OR Tambo very poor.

The next trip was with Air France on their A340-600, then a couple of times with Emirates on their Airbus A380.

We flew on the A380 from London to Dubai and sadly transferred to the B777 to Johannesburg.

This was my first trip on the A380, and I was hooked, not just on the plane, but with the service with Emirates.

During our many trips, Helen got to see much of the country from the borders of Namibia through much of the Cape province, Gauteng, KZN and Mpumalanga.

Some of the highlights were white river rafting on the Orange river between Namibia and South Africa, trips to the Kruger game park, a game park in KZN, a visit to the Cradle of Mankind, walking hand in trunk with an elephant, stroking a cheetah and a visit to Robin Island.

When we visited the Island our tour guide allowed me to actually go inside Mandela's cell.

Tourists were not allowed in the cell. They could walk past and take pictures, but that's as close as they got to the cell.

For once I beat Michelle Obama. A few years later she too was allowed to enter Mandela's cell. Obviously it was only open to VIPs.

Another highlight for me was visiting a sanctuary for injured animals. I got to feed a vulture after it swooped down and landed briefly on my arm to snatch the piece of meat.

Helen also did a quad bike ride in another game reserve passing close to some rhinos.

For someone who only weeks before got her driver's licence and then had to contend with the lawless and reckless South African

Travelling by sound and smell

drivers, the state of the roads and various signs warning motorists of crime hot spots and high jacking areas, she deserved a medal.

To this day I feel guilty that she had to do all the driving.

South Africa is the country of carnivores and I don't mean the animals, but the people.

It is a meat eating lovers heaven.

After arriving in the country, I couldn't wait to show Helen what T-bone and fillet steaks were all about. They were nothing like the pathetic tiny cuts of meat available in the UK.

When we visited our first supermarket, she couldn't believe the size of cuts of meat.

South Africans love to "braai", the Afrikaans word for a barbecue, and I don't mean using gas, but the real thing, charcoal or even better some form of hard wood.

Meat served at the family braai is often accompanied by "Pap" and is a staple food throughout Southern Africa.

Pap is similar to polenta.

How To make Mielie Pap which is simply stiff porridge made with fine maize meal (cornmeal), adding water and a pinch of salt.

You will need 500 ml water, about 200 gm maize meal or corn meal and a pinch of salt.

Heat the pot with water over medium heat until lukewarm. Add about half of maize meal and a pinch of salt and stir well.

Keep stirring with a wooden spoon until the maize meal mixture begins to boil and bubble. Let it cook for about 2 more minutes.

Next slowly add the remaining maize meal with the left hand while stirring with the right hand. Continue stirring until the mixture is cooked throughout.

You might need to add more water, but you don't want the mixture to be too runny.

I had previously made "pap" for Helen and her friends.

My alternative recipe included frying some smoked bacon bits, mushrooms, garlic and adding a tin of sweet corn to the "pap" before sprinkling some cheese over the top and baking it in the oven for a few minutes before serving.

If you are going to make "pap", a handy hint is to immediately after cooking, before serving, fill the bottom of the pot with water to let soak and clean the sides of the pot. Clean the wooden spoon too. Because once the porridge has hardened, it's very hard to clean off.

Most indigenous folk serve the basic "pap" with a stew.

Lamb sosaties are also a common feature of the classic South African braai. The apricot and curry marinade might sound unusual but it adds a wonderful flavour to these tasty kebabs.

Sosaties are a Cape Malay dish and the name comes from the words "sate" meaning skewered meat and "saus", a spicy sauce in Afrikaans. While this is a dish that originated in the Cape Malay area, it has spread in popularity around the country.

There are a few variations to the sosatie marinade, but most at least include apricot jam, curry powder and garlic. Most recipes I found use wine vinegar, although I have seen some mention of tamarind which I imagine would be a tasty alternative.

This marinade can be used with beef and chicken as well. Lamb is simply the most common way to make these skewers, but the apricots do go particularly well with lamb.

HOW TO MAKE SOSATIES

Soften the onion in a little oil and add the garlic and ginger.

Add the vinegar, jam, curry powder and cumin. Mix

Travelling by sound and smell

well then leave to cool.

Dice the lamb then mix through the cooled marinade.

Leave a few hours or overnight.

Soak the dried apricots in warm water to soften.

Thread the meat onto skewers along with dried apricots and chunks of onion.

Cook until the meat is cooked through then serve.

If you are using onion, you want a softened onion rather than raw onion.

You can also try using different veg and fruit.

One of the great things about these sosaties is they can be prepared ahead and then they take just a few minutes to thread on skewers, cook and enjoy.

I also wanted her to try Cape Malay mutton curry. When the dish arrived at our table, she was somewhat taken aback to discover the curry included the meat on the bones.

After getting over the shock of bones in curry, it didn't take her long to begin to enjoy the various South African delicacies.

The first South African delicacy was biltong.

This is sticks of dried and cured beef either sliced instore, or at home. It can also be ostrich meat.

Biltong is not the same as the American jerky. Jerky is thinly sliced beat placed on a rack and cooked at a slow and low temperature, while biltong is cured and marinated for 24 hours then hung on hooks to air dry for up to a week depending on how dry or wet you want your sticks of biltong. Once ready, you can slice it or get the local butcher to do this.

Beef is the most common form of biltong but game including springbok, impala kudu and other buck and even elephant biltong can be found. The only difference with game biltong is it needs to be extra dry.

The best way to find out if game biltong is ready to eat is to see if you can snap the biltong stick in half. That's how dry it should be.

Now let's get back to the curry.

The most important thing to remember about Cape Malay curry is that the spices are warming, sweet and earthy but don't blow the roof off your mouth.

The meat is cooked in a thick gravy of onions, tomato and spices and potato is added near the end.

Slow cooking is required here, to get fork tender meat and potatoes. Rapid cooking doesn't give the same taste or texture to the curry.

Curry leaves give a lovely aroma but can be left out if not easily available without any change in the end result.

A handful of chopped coriander added before serving gives a fresh taste.

No fancy spices like fenugreek or garam masala are required.

How to make Cape Malay curry
Ingredients
60ml sunflower oil,
400 grams
about 2 medium onions, finely chopped,
140 grams tomato, finely chopped or grated,

Travelling by sound and smell

5ml granulated sugar,
2 cinnamon sticks,
4 cardamom pods, bruised,
4 allspice berries,
4 cloves,
30 ml crushed garlic,
30 ml grated ginger root,
5 ml turmeric powder,
7.5 ml red chilli powder,
5 ml chilli flakes,
7.5 ml cumin powder,
7.5 ml coriander powder,
5 curry leaves
1 kg stewing mutton pieces, rinsed and patted dry,
5 ml salt adjust to taste,
500 grams potatoes, peeled and cut into quarters,
60 ml hot water
30 ml coriander leaves chopped.

Method

Heat the oil in a pot over medium heat and add the chopped onions. Sauté until translucent, about 15-20 minutes.

Add the chopped and grated tomato and cook together until the onions and tomato look golden, for about another 10 minutes.

Add all the sugar, all the spices and the water and simmer together for 8-10 minutes to combine the flavours and cook out the spices. The water prevents the spices from burning. Add the stewing lamb and salt and mix together to cover all the meat in the spice mix. Replace the lid on the pot and simmer on low for 30 minutes.

Add the potatoes and stir through so that they get covered in the curry sauce. Cook for another 25-35 minutes until the meat and potatoes are fork tender but not mushy.

Check the seasoning and add more salt if required.

Add the chopped coriander and give it a final mix through.

Serve hot with roti, naan or fragrant basmati rice.

There is also what many call Durban curry, which is very different in taste to the Cape Malay curry.
 Another favourite South African dish Helen got to try was Bobotie: A classic South African casserole.
 It is made with ground beef, seasoned with a creamy topping. It is rich, savoury, spicy and aromatic.
 Bobotie is a South African casserole with curried ground beef at the bottom and a thin layer of egg custard on top.

A bit of bobotie history.
Bobotie's roots in South Africa date back to the 17th century. Dutch traders set up camp in the area now Cape Town as a stopping point on their journeys back and forth to Indonesia. The

Travelling by sound and smell

traders brought spices, cooking techniques, and recipes with them.

While the specifics are a bit vague, it is thought by some that the original bobotie recipe came from Indonesia and was adapted to fit the available ingredients.

Today many consider bobotie to be the national dish of South Africa, and it has become popular on menus featuring South African cuisine all over the world.

How to make babotie
Ingredients
3 slices of bread,
1½ cup milk (divided)
2 tablespoons olive oil,
2 large onions roughly chopped,
4 teaspoons medium curry powder,
1 teaspoon dried herbs (whatever you've got around - oregano, basil, marjoram, etc.),
1 teaspoon ground cumin,
½ teaspoon ground turmeric,
2 garlic cloves finely chopped,
1½ pound ground beef,
½ a cup of fruit chutney,
1 tablespoon apricot jam,
Zest and juice of one medium lemon divided,
4 teaspoons tomato paste,
Salt and pepper,
2 large eggs,
4 bay leaves.

John CT Miller

Method

Preheat your oven to 350°F.

Soak the bread in 1 cup of milk.

Heat the olive oil in a wide skillet set over medium heat. Once the oil is hot, add the onions, and cook until soft.

Add the curry powder, mixed herbs, ground cumin, turmeric, and garlic, and stirring constantly, allow to cook for a minute or two until the garlic is soft.

Add the ground beef, and cook, stirring frequently to break up any big chunks, until browned.

Once the beef is browned, remove the skillet from the heat, and stir in the chutney, apricot jam, all the lemon zest, half the lemon juice, tomato paste, and salt and pepper to taste.

Mix well, then give it a taste, and if needed add more lemon juice, salt, and pepper.

Next squeeze the milk from the bread, reserving the milk for later, and tear the bread into small pieces. Mix the bread into the beef mixture, and spread evenly into an oven-proof dish.

Strain the milk that has come from the bread, and add

the remaining 1/2 cup of milk. Beat in the eggs, and season with 1/4 teaspoon salt and pepper to taste.

Pour this over the ground meat, and scatter the bay leaves on top.

Bake, uncovered, at 350°F for 45 minutes, or until golden brown.

In my opinion, it is the curry powder and the bay leaves that influence the flavour of this dish so use the best quality that you can.

I like to prepare and bake this recipe in a cast iron skillet - it can go from stovetop to oven, and even looks beautiful on the table. Plus it makes for fewer dishes to wash later!

Before finishing my brief recipe guide I have to add a couple of desserts.
Malfa pudding is probably one of the best known desserts.
It all starts with a wonderful spongy cake, flavoured with apricot jam, that bakes until the exterior becomes somewhat caramelized. Then, after poking holes throughout the sponge, a mixture of heavy cream, milk, butter, sugar, and salt, is poured over the pudding. The textures of the caramelized exterior, spongy centre, and delicious creamy liquid are definitely Moorish.
I don't know if it's possible to make a cake that is richer or moister!

John CT Miller

How to make malva pudding

8-inch (20cm) square baking dish, Mixing bowls, Saucepan, Two bowls of Malva Pudding, served with vanilla ice cream on top.

Preheat your oven to 350°F (180°C) and butter an 8-inch (20cm) square baking dish.

In a medium-sized mixing bowl, whisk together the milk, brown sugar, eggs, apricot jam, melted butter, and vinegar until they are thoroughly combined.

In another mixing bowl, whisk together the dry ingredients, the flour, baking powder, baking soda, and salt, and then combine the dry with the wet ingredients until thoroughly mixed.

Pour the mixture into your prepared baking pan and bake for about 30-40 minutes until you can insert a knife and it comes out clean.

While the pudding is nearly finished baking, make your sauce. In a saucepan over medium heat, combine the cream, milk, butter, sugar, and salt. Heat until the butter is melted and the sugar is dissolved.

Once the pudding is finished baking, poke holes all over the hot pudding with a skewer and then pour the warm sauce over the pudding. Let the pudding rest for at least 30 minutes before serving warm.

Travelling by sound and smell

Another South African favourite found in speciality shops and local markets are koeksisters.

Koeksisters are "A very sticky South African sweet. They remind me of a donut but much sweeter.

There is also a Cape Malay version, but these are mainly found in the Cape Province.

To make the Syrup you will need 250 ml water, 625 ml white sugar, 12 ml lemon juice and 5 ml vanilla essence.

For the Dough you will need 375 ml cake flour, 22 ml baking powder, 1 ml salt, 20 grams butter, 150 ml milk (vanilla soy milk gives added flavour and 750 ml canola oil.

Method

Put the water and sugar in a pot and bring to boil on low heat. Stir frequently until the sugar is completely dissolved. Boil for 7 minutes.

Remove the pot from the stove and stir in the lemon juice and vanilla essence. Put the pot into the fridge.

Mix the flour, salt, and baking powder thoroughly in a mixing bowl. Break the butter or margarine into small pieces and add to the mixture. Add the milk. Mix well until a dough is formed.

Roll the dough out to a thickness of 5 mm (+ or - 1/4 in.). Cut the dough into thin (+ or - 10 mm or 1/2 in.) strips. Take 3 strips and join their ends on one side. Braid the strips to desired koeksister length and join other ends.

Heat the oil in a pot until very hot. Put about 3 koeksisters (or what can fit)at a time in the oil and fry them on both sides until they get a golden-brown colour. As you remove the koeksisters from the oil, dip them directly into the syrup you prepared and placed in the fridge.

Remove the koeksisters from the syrup and allow the excess syrup to drip off.

Place them in the refrigerator to cool and then eat!

It is important to keep the syrup cold, so between dipping the koeksisters return the syrup to the fridge to maintain its coolness.

You can read more about our trips in my book (Fishing for Hyenas and looking for Sleeping Penguins).

After our fifth trip to South Africa, we went our separate ways.

Thanks for the time together Helen, and all you had to put up with.

Chapter 27

They say when one door closes, another one opens. In my case, it was another cabin door.

But before we take to the skies, at this point I think I should return to the terminal and explain what happens to disabled and often elderly passengers when travelling alone.

I am not sure when it began, but at every airport I visited wheelchairs became the order of the day.

As far as I remember, it started even when I was with Helen.

It didn't seem to matter if I was with someone or not, ground crew around the world insisted on me sitting in a wheelchair.

In most cases, this seemed to be non-negotiable.

With Helen or my guide walking next to me as I sat in the wheelchair, we made our way through to the baggage carousel after which the wheelchair most times would be discarded.

Stepping out the aircraft, it was sometimes a woman pushing a wheelchair.

It often made me feel extremely uncomfortable. I often said "I might be blind, but not cripple".

Some ground crew understood my protestations, but said they had been instructed to push me in the wheelchair through the airport terminal.

Something else which annoyed me about travelling on my own and confined to a wheelchair was been taken straight from the aircraft to the departure area to wait at times up to a few hours for the next flight.

This is fine for someone sighted, but when you are blind, you have no idea where you are, and also more importantly where the toilets are situated.

What would happen if the blind person needed the toilet?

After a couple of hours the man or the woman responsible for pushing the wheelchair would return, and push me to the doors of the plane.

I do remember once at Suvarnabhumi airport in Bangkok being left even though I told the person if they did this I would get out of the wheelchair and ask fellow passengers to take me to where I wanted to go.

On that particular occasion having been left somewhere, and after walking into various objects I finally found my way to an information desk.

A somewhat heated debate followed, and they finally found someone to take me to where I wanted to go.

This as far as I was concerned, is blatant discrimination.

On one of my trips with KLM after the B777 landed in Taipei Taoyuan International Airport the wheelchair man refused to take me to duty free.

Fortunately, his refusal happened shortly after I sat in the wheelchair.

I told him if he didn't take me there, I was going to get out of the wheelchair and find my own way to duty free.

At that stage the head of the passenger assistance department was nearby and asked me what was the problem?

When I told her he had refused to take me to duty free, she spoke rapidly to him, and following her intervention, he then took me where ever I wanted to go.

Another thing which annoyed me about being wheelchair bound

Travelling by sound and smell

or someone who needed passenger assistance was to be always taken on board first, and always last to exit the plane.

Each time I was taken to the aircraft doors by wheelchair, one of the cabin crew would guide me to my seat.

The worst airports for passenger assistance and been treated like cattle are Heathrow and Gatwick.

After clearing customs and security at these airports, I would be taken to a holding pen as such until it was time to board.

Some hours after waiting in the holding pen, a several seater electric buggy arrived and we would all be taken to the appropriate gates and once again waited until it was time to board the aircraft.

This herding like cattle also happened each time returning to London.

While on the subjects of airports, I have to mention the following experiences.

The most unfriendly airport has to be Ataturk or the new international airport in Istanbul. On arrival one day, the ground staff refused to take me anywhere, and left me at an unmanned office and disappeared until it was time to board my next flight.

I had no idea where I was. I think it was outside an office, but there were certainly no passengers or the sound of voices nearby.

It always amuses me as soon as you challenge workers, all of a sudden they cannot understand you or speak English.

Mind you Turkish Airlines was also one of the worst airlines I travelled on. The cabin crew did not speak English very well, and like their companions on the ground were not very friendly.

After my one and only flight with Turkish Airlines on their B777, I vowed never to fly with them again, even if they offered some amazing deals.

One of the few positive memories about this airline was their fruit juices: Sour cherry and pomegranate.

The other was being able to fly on both the airbus A321 and the A319 to and from London.

Another airline which I vowed never to fly with again was the highly rated Cathay Pacific.

At the time the only reason to fly with this airline was because they were using the Airbus A350. I had never flown on this aircraft before.

The cabin crew were not very friendly, and also didn't seem to understand English very well.

Once guided to my seat, flight attendants were always surprised when I refused to be seated.

When booking my flights, I always chose an aisle seat.

Most times if the disabled passenger like me chooses an aisle seat, it meant having to stand up when other passengers wanted to take their seats next to you.

So what's the point of been taken on board first?

I always had to explain to the flight attendant I was not going to take my seat until everyone was on board and seated.

Having refused to be seated, I always made my way to the galley and waited for all the remaining passengers to board.

No matter which airline, the cabin crew in the galley would offer me something to drink, but not Cathay Pacific.

While waiting to take off from Heathrow, they insisted they were not even allowed to offer me a glass of water, let alone fruit juice.

The food was also pretty poor.

When I asked them about what movies were available and if they included audio description, they didn't seem to understand.

For years I had heard what a great airline Cathay Pacific was, but like Turkish Airlines, I will never fly with them again.

The ground staff in Hong Kong were also not that much better.

Having dealt with the 2 worst airlines, let me tell you about the

Travelling by sound and smell

best carrier.

I can't remember how many times I have flown with Emirates, but the cabin crew were always fantastic. They could never do enough.

Another plus for Emirates was once on board the A380, passengers were offered a choice of no less than 12 spoken languages. Compare this to other airlines which offered a choice of 2 languages if you were lucky.

While singing the praises of Emirates, I cannot forget the ground staff at Dubai. They were just as friendly and helpful. They would always take me to wherever I wanted to go.

Qatar airways is also a great airline, but still doesn't beat Emirates.

The flight attendants are also very friendly and just like the ground staff at Dubai, those in Dohar are just as helpful.

Most of the ground staff at these airports were manned by Filipinos, Indians or people from Bangladesh.

One middle east airline which did not follow the example set by Emirates and Qatar airways is Kuwait airways. On one of my flights to Manila on the B777 and back again on the B767,

On board service was poor and shoddy. However, the ground staff at Kuwait city were a lot better.

On the other hand, one airline which was pretty efficient and helpful, was Thai Airways, but still not as good as Emirates and Qatar Airways.

The ground staff and officials at Suvarnabhumi airport in Bangkok were most times friendly and helpful.

Another airline I often flew with was the Philippines low-cost airline Cebu Pacific.

They were much like their colleagues at Ninoy Aquino airport in Manila always cheerful and helpful.

While the ground crew at tan Son Nhat airport in Saigon were always obliging, they did not speak English very well, but they tried their best and always went out of their way to help.

One of the many good things about this airport, passengers were able to prepay the set taxi fare before leaving the terminal.

Once the new arrival received the receipt, they simply walked outside and hailed any of the waiting taxis to take them into the city.

This scheme meant passengers could never be ripped-off by unscrupulous taxi drivers.

I wish airport authorities around the world followed the Vietnamese example.

How many times have passengers been ripped-off by crooked taxi drivers?

My only stopover in Kuala Lumpur in Malaysia flying from Saigon to Manila one day lead to a major confrontation with a wheelchair man.

Once all the passengers had disembarked from the Air Asia B737, it was my time to leave the plane.

Exiting the plane the wheelchair man was already waiting for me.

He told me to sit in it.

I said I was not going to sit in it.

He was adamant insisting I should sit in it.

I told him I had not ordered the wheelchair and had no intention of sitting in it.

After a moments silence, he then said I should pay him for the wheelchair.

As you can imagine, I told him to leave with the last word being off, and take his wheelchair with him.

Following his departure, my guide and I then made our way through the airport to find our connecting flight to Manila.

Travelling by sound and smell

Out of all the airports I have visited Ninoy Aquino airport in Manila and OR Tambo in Johannesburg have the friendliest and most helpful ground crews.

There was never an issue when I told them I wanted to rather walk, than be pushed in a wheelchair.

They also always took me to where ever I wanted to go in the airport.

Most times it was first to duty free, and then on to a coffee shop where they left me until it was time to board.

While on the subject of OR Tambo, I must relate the following 2 stories.

On one of my flights between Durban and Johannesburg on the B737 with the now defunct Kulula airways, I actually managed to disembark before anyone else.

On this flight, it is still one of the only times I exited the plane before anyone else.

I told the concerned flight attendants if they didn't let me get off first, I would simply walk out and find someone to help me.

Prior to that flight I had contacted a blind black lady in Johannesburg after hearing her talk on a local radio station about how she was unable to use her cell phone to connect with friends and family.

After hearing her predicament, I called the producer of the show and managed to get the number for Nosisa.

Some of you might not know, but the iPhone is the only cell phone blind people can use without the help of someone sighted.

I promised her that on my next trip to South Africa, I would buy and bring her an iPhone.

We agreed to meet along with her daughter at OR Tambo.

However that day, the flight from Durban to Johannesburg was delayed, and to make matters worse, I had to get a connecting flight back to London.

Fortunately, we landed at OR Tambo about 30 minutes late and I knew I did not have time on my side, and also I knew Nosisa would be waiting for me.

Well, after exiting the aircraft, I managed to find Nosisa, who was probably more worried than I was.

Over a cup of coffee, I gave her a 10 minute rundown on how to use the phone, before we parted company.

Here is a copy of the article which was written about my meeting with Nosisa.

"Blind Date at O.R. Tambo International

A chance late night call from a blind Vosloorus woman to Talk Radio 702 resonated with a man in Wales, who decided to make her life a little better. With the connivance of a 702-call screener, contact was made.

Emails went back and forth for the next six months until he and she met at OR Tambo International, earlier this year. No, this is not a love story; it is a feel good tale of a man who, for sentimental reasons, listens to South African radio stations.

John and Nosisa

John Miller is the man in Wales. Dubbed "John from the UK" by 702, he's a conservation writer, who also specialises in travel for the blind. Unbeknown to the anchors of the radio stations he calls, John is himself blind. Here is his story.

"As a blind person, radio is my preferred medium for news and chat. I spent my formative years in Zimbabwe and most of my working life in Johannesburg, South Africa, which probably explains why Radio 702 is on my radar.

Late night listeners would know me as an active caller-in. Early one morning in the middle of last year, I tuned in to Gushwell Brooks who was talking to a blind woman referred to as "Nosisa from Vosloorus."

Travelling by sound and smell

She was recounting a conversation she'd had with her teenage daughter, Zandile, who wanted to know how her mother knew she was beautiful.

Nosisa replied, "My daughter, I don't have to be able to see you, I can see you in my heart, and my heart tells me you are beautiful."

Those words touched me and I slowly built up a picture of Nosisa through her regular calls to the radio station. I discovered she was unable to use social media as she would have liked, because her 8-year-old phone was unable to cope with the latest applications.

Unbeknown to me at the time, Nosisa's husband Kubuse is also blind, but it was he who he acted as the go-between for the next several months; we exchanged emails and I sent hints and tips about how a blind person could use an iPhone.

The months passed and a long-planned trip to visit my brother in Durban finally arrived.

Ahead of my return journey, I phoned Nosisa from Durban – it was the first time we actually spoke to each other.

We arranged to meet at Wimpy while I was in transit at OR Tambo International in Johannesburg.

As Murphies law would have it, my Kulula flight was late.

Airlines throughout the world have a policy where disabled people or those who require passenger assistance are first on and last off.

On this occasion, I told the cabin crew I was not going to wait to be last out and if need be, I would make my way on my own out of the aircraft.

This urgency was necessary as Nosisa had called me twice while waiting for the doors of the aircraft to open, probably still wondering if I was going to meet her, and probably more importantly if I had the promised iPhone.

The crew obliged me and soon I was greeted by two of the

ground staff. I impressed upon them my need to go ASAP to a meeting with a blind lady.

Nonetheless, Murphies law stalled me again and my luggage seemed to be the last off the carousel.

At last I made it to my destination, where a much-relieved Nosisa awaited me. There were hugs all round and then, against the background of a clatter of plates, knives and forks and chatter and laughter from nearby tables in the restaurant, it was time to begin my lesson on how to use an iPhone.

The drum rolls were imaginary as I handed the Steve Jobs' invention to Nosisa and a few minutes of silence followed.

Nosisa later described how those first few moments were filled with wonderment, disbelief and gratitude. So many questions filled her head: "Was this really happening?

Had it really happened? Am I really holding the promised phone? Is this really John?"

Amid the mixed emotions we all felt, I gave Nosisa an idiot's guide (I'm no expert) to how iPhone can help blind people gain their independence.

I hope by now she has joined those of us reading and sending messages via text and email, using WhatsApp and Skype and communicating with the outside world like the rest of the planet.

I cannot begin to tell you the sense of freedom this gives a blind person.

Blind people are no longer excluded from the world of social media, nor do they have to rely on someone sighted to read or write for them.

I have immense gratitude for the myriad dedicated apps for blind iPhone users, which can in real time tell the user what colour his or her clothes are or what colour objects are around them.

There are apps to read bar codes and labels, to let us know

whether lights are switched on or off and to assist with foreign currencies.

There is even an app to reveal whether someone is smiling at you!

In short, the world is now Nosisa's oyster or put another way, Nosisa is now part of the real world.

Perhaps next time you upgrade your iPhone (providing it is one of the 8 series or later models), you too can bring light and joy to a blind person's life."

This article first appeared on Suzanne's World, well known South African writer Suzanne Brenner's newly launched digital space for stories and snippets she likes, which sits alongside her business website **PROWRITE WRITING SERVICES**. See the original here. Suzanne says: "John Miller's feel-good story has not only inspired others, but it has motivated him to continue being a Good Samaritan, which I'm sure he'll share with Suzanne's World."

Suzanne hopes the new section will afford her the luxury of nurturing new writers who love writing.

* Without the help of 702 personnel, this meeting would not have taken place. Thanks to presenters Gushwell, Florence and call screener Mmakgomo, who made the meeting possible.

iPhone changes lives

Apple founder Steve Jobs and his designer successor, Sir Jonathan Ive, were the first to make smart phones user-friendly for blind people, thanks to a built-in voice or screen reader called VoiceOver. VoiceOver is among Apple's Accessibility features where the iPhone (4S and upwards) reads out the text on the screen.

With a series of double taps on each of the icons or applications and by using one, two or three finger swipes, the phone loudly vocalises exact locations on the screen, thus enabling the blind person's ability to navigate the way through the various pages of applications.

To activate VoiceOver, go to settings, then general, then Accessibility and turn VoiceOver ON.

Siri, Apple's voice-controlled personal assistant, also enables a blind person to send and receive SMS messages.

The other memorable flight into OR Tambo was with Turkish Airlines or rather one of its passengers.

On that flight I was seated next to a South African woman with a pseudo American accent. She had apparently moved to the states a couple of years earlier and married a body builder.

The marriage didn't last long, and after hearing her entire life story, and we landed, we went our separate ways.

We had spent the whole night talking. She was on her way back to spend time with her family, before returning to America.

In the months and years that followed, we kept in contact, or rather she kept in contact with me and each time she called, I couldn't believe the various schemes she got involved with.

If ever there was an adventurous person in life or someone with the gift of the gab it was this lady.

At some stage she met an Egyptian man, and decided to get married to him and move to Cairo.

When she left America, she did not want to leave her pet dog behind, so pretended it was one of the many assistive animals, which allowed her to travel with her dog.

How she fooled the airline, I will never know.

After a short while in Egypt, the marriage apparently didn't last long and she then, who knows how took on a job teaching English.

At first giving private English lessons, but soon after managed to get a post at one of the schools.

How she conned her way to do this, I will never know.

Another memorable flight for completely different reasons was on a KLM B737 to Amsterdam.

I was seated next to a 10 year old girl with her mom.

One thing I have discovered and appreciated about most young people they do not have any inhibitions or feel embarrassed about asking direct questions.

Most adults are invariably dying to know how I was blinded, but for whatever reason are too inhibited to ask, while young people do not hold back.

This young girl wanted to know all about me, how I went blind, and went ahead with numerous other personal questions.

In between answering questions, I managed to find out from the mom, she was taking her daughter on a 2 day whatever she wanted to buy birthday trip.

When we deplaned, the daughter insisted on holding my hand and guiding me to the next departure point.

It's definitely the little things in life that count.

Chapter 28

Before I leave my thoughts of airports, airlines, aircrafts and passengers I must mention the following.

Part of the world wide safety regulations when it comes to blind passengers is for the flight attendants to explain to the blind passenger where everything is and how things work.

As a former aviation journalist, I obviously knew the procedure, but kept this to myself after each time boarding an aircraft.

The cabin crew is supposed to tell or show the blind person where the toilets are, tell them where the emergency exits are, show them how the oxygen mask works and where the life jacket is kept and where to find the call me button.

It is probably necessary that blind passengers flying for the first time are shown how the seatbelt clips in.

They are also supposed to take the fold-up cane away for take-off and landing and place it in the overhead locker.

Sadly many cabin crew don't do any of the above.

I normally waited about 2 or 3 hours into the flight and then requested to speak to the senior purser.

I would then ask why the cabin crew had not told or shown the blind passenger which is me, where and how everything worked?

This complaint is most times greeted with numerous apologies with VIP treatment to follow for the rest of the flight.

However, I always pointed out to the purser, my complaint is not about me, but my concern for other blind passengers who do not know what the rules and regulations are.

Travelling by sound and smell

Many airlines began each flight by proudly offering me a copy of their safety pamphlet in Braille.

However, this pamphlet like most braille documents is bulky and cumbersome and not that easy to read even if you were a good braille reader. I am not a fast or even good braille reader.

I always suggested to the flight attendants that instead of carrying a braille safety pamphlet on every flight, it would cost a lot less to have the safety pamphlet recorded in at least 2 languages on one of the on board entertainment audio channels.

I somehow doubt if my suggestion has ever reached management.

If only airlines and most other commercial companies actually asked blind people what would make life a lot easier, but they seldom do.

Now with cabin rules and regulations explained let's get back to on-board entertainment.

On one of my previous flights with Emirates on their B777 I was excited to learn the airline had introduced live TV news and sport followed shortly after that by Wi-Fi.

Unfortunately these features at the time were not available on all its aircraft.

For whatever reason the services seemed to be first rolled out on their B777 fleet.

When I was told about these features, I couldn't wait to listen to the live news channels.

Before the pandemic, Emirates offered their passengers access to eight different live news and sports channels, including BBC World News, CNBC and CNN International and a couple of sports channels.

However the channels along with Wi-Fi and most of their flights were all suspended during the pandemic.

Live sports are back again, and will be shown across two channels on Emirates' in-flight entertainment system.

Sport 24 and Sport 24 Extra are the world's first and only live sports channels for the airline industry and Emirates have the rights to show nearly all the big tournaments including Premier League soccer matches, NFL, Ryder Cup, and more.

Sport 24 broadcasts 24-hours a day with around 16 hours of premium sports content per day, while Sport 24 Extra broadcasts around 150 hours of live sports entertainment per month.

At the time of writing this, I believe the airline has almost 200 aircraft broadcasting live sport and offering over 5000 inflight channels.

It was on the B777 route from Dubai to Manila the airline had also introduced Wi-Fi.

At that stage, the internet was not available on all their aircraft.

In the beginning, passengers paid $1 for 500mb. However, it didn't take long before the $1 charge went to $10.

Having paid the $1 signing up fee, I called a friend in the Philippines to show her what the food looked like on my tray.

She was completely surprised to hear from me so soon after leaving her country.

About 3 years before the pandemic and apart from travelling at 30000 feet, something else I looked forward to was catching up or listening to a few of the latest movies.

I don't know what took me so long to get round to on board movies or documentaries, but for most of my long distance flights, I often listen to one of the books on my Kindle or my iPhone.

Now many of you will probably not know, but most movies these days also include an audio described track which can be enabled while watching the movie.

This track tells the blind person what is happening at times and

Travelling by sound and smell

describes the surrounding scene.

For example the track would say as a man or woman enters a shop, the audio track will say and describe the person, their actions, facial features, running or walking. I hope you get the idea?

It didn't take long to discover that most cabin crew did not know about this audio described feature.

In fact the Cathay Pacific cabin crew did not even know how to find the movie section. Hardly surprising, they also didn't bother to let me know where the exit doors were or show me where the oxygen mask and life jacket were stored.

When I later reminded them of their failure to show me where things were and how they worked, they simply said they were sorry, but that's as far as it went.

I got the feeling they weren't sorry at all. It was in one ear and out the other. They just didn't care.

Chapter 29

The first of my many trips to South East Asia was more by luck than desire or planning.

If I am honest, the Philippines was never on my must visit map. The only thing I knew about the country was the Beatles played 2 concerts in Manila.

My knowledge of Beatles history paid off one day when I won a competition to fly to Manila in the Philippines where the Beatles played a couple of concerts in July 1966 before they were thrown out the country.

I have been a lifetime Beatles fan from the first time I heard their song (Love me do).

Thanks to Youtube I have managed to listen to many of their concerts around the world.

The only concert I have never found online was the Beatles in Manila.

When the Beatles arrived in Manila they were taken to a navy base to do a press conference. I wouldn't be at all surprised if this was the only time a naval base was used for a meet and greet.

Following the typically no doubt boring and predictable questions, the group were taken to a luxury yacht on Manila bay to spend the night away from the thousands of screaming fans.

However, after a few hours, the band insisted on returning back to dry land and to a hotel, where they spent the next 2 nights.

Thanks to their late manager Brian Epstein, the Beatles had decided never to get involved in promoting any politicians.

Travelling by sound and smell

Imelda Marcos the president's wife however had other plans. She had organised a breakfast at the palace hoping the Beatles would attend and even play a few songs for her.

All the well-connected were invited that morning and in one of the many palace rooms even the national symphony orchestra performed, but the Beatles remained at the hotel.

The first lady was not used to being rebuffed, and within hours the situation for the group changed.

When Imelda realised she had been snubbed, the security at the hotel suddenly was withdrawn. Even room service seemed to disappear.

With the lack of security, a few teenage girls managed to go upstairs and a few actually met 3 of the Beatles.

After the incident with the first lady, the Marcos controlled media turned on the Beatles and once again at the airport, there was no security and the band and some of the touring party were roughed up before boarding the plane.

The Beatles couldn't wait to get out the country, and this incident in that year probably lead to the band deciding they would never tour again.

Now back to my trip.

The flight with KLM took me from London on a B737 to Amsterdam and then on via Taipei to Manila on a B777.

When we landed at Taoyuan airport in Taipei the obligatory wheelchair was there to meet me.

I told the man pushing the wheelchair I wanted to visit duty free, but he insisted this was not possible and wanted to take me to the next boarding gate.

I realise he was probably only doing his job, but I immediately got out the wheelchair and said I was not going a step further.

I called for the head of the ground staff and explained the

problem. After listening to me, she made it clear to the man even though I could not understand her commands, he was now ordered to take me where ever I wanted to go in the airport.

After the refuelling stopover, we continued our flight on the B777 to Ninoy Aquino airport in Manila.

Leaving the B777 in Manila, the wheelchair was once again there to meet me.

This was my first time I had met a Filipino, and as things turned out, he would be one of many during the next few years.

I even got to know a few of the ground crew as a result of my regular visits to Manila.

They were always incredibly obliging and went out of their way to help me.

Chapter 30

Before leaving for Manila, I had done a lot of online research about the Philippines and Manila and discovered there were a few places I wanted to visit.

The Philippines consists of over 7000 islands with Luzon and Mindanao islands making up 65% of the land mass.

Manila was originally founded by the Spanish in 1571 and in 1898 the Spanish seeded Manila to America. Today metro manila comprises 16 districts.

The population of Manila is well over 20 million.

In 2019 it was recognised as the most densely populated city in the world and also the second most natural disaster afflicted city in the world after Tokyo.

Thanks or no thanks to America, the Philippines is also linked to spam, the salted and processed pork in a can.

When America rescued the Philippines from the invading Japanese during world war 2, its military handed out thousands of cans of spam to the starving crowds in the country.

Spam was originally invented in 1937, and soon became a standard ration for the military.

No matter where the American military went, spam went with them.

After world War 2, many Filipinos settled in America, and began sending back boxes of span to relatives back in their homeland.

Believe it or not, there is even a spam museum in America.

It really makes no difference how much research one does about a foreign country, and how prepared you think you are, you really need to be there to find out what the country is really like.

For me, it was all about the sounds, smells and gastronomy of each city and country.

I don't know what it is with taxi drivers outside airports, but many are con artists, and often take innocent visitors for more than just a ride. Manila was no different.

The first taxi driver I spoke to wanted to overcharge, until I got my airport assistant to take me to another taxi driver.

Eventually after telling a few taxi drivers they were crooks and thieves, I managed to get a driver to take me to Hotel Artina in the CBD.

I later discovered even though I thought I had got a good deal with the taxi driver, he had still managed to rip me off.

Oh well, at least I knew how much to pay in the future.

Arriving more than one hour later at the hotel, I was amazed at the welcome I got from the staff at the hotel.

This was their first time to meet a blind person. Unlike many other people around the world, it was almost as if they had been dealing with blind people their entire lives.

There were never any awkward moments and many were happy to be seen out on the streets with me.

In the years that followed, I got to know all the staff and all about their families.

One of the receptionists Aileen became a special friend, and later I would always visit her family at home each time before I left the country.

Before leaving for Manila, I had also managed to find a guide to take me to the various places on my must visit list.

Obviously first was a visit to the national stadium where the Beatles played.

Travelling by sound and smell

Visiting an empty stadium certainly goes nowhere to providing an atmosphere, but at least I was able to say I had been there and got the t-shirt.

Believe it or not those tickets for the Beatle's afternoon and evening concerts at the national sports stadium cost less than $1, and there were even tickets for 50 cent and even 20 cents.

Another memorable moment was a trip to lake Taal where I made my way to the top of the active volcano situated in the middle of the lake.

This was going to be my first close up experience of a volcano and not just any old volcano, but an active volcano.

As we made our way in a boat across the lake to reach the volcano, the smell of sulphur got even stronger.

After leaving the boat, I decided to hitch a ride on horseback. Making our way up the 3 kilometres to the rim of the volcano, I could not believe how hot some of the rocks and ground were.

Standing on the rim didn't do much for me. My thoughts at the time were how peaceful everything seemed and sounded at that moment.

No doubt when it erupted, I guess the silence would be shattered.

Many years later, the volcano did in fact erupt sending clouds of ash across the nearby land.

Somewhere else I visited was the main cathedral in Manila. This was where Imelda Marcos used to worship. According to my guide, it was the only church in the whole of the Philippines which had air conditioning installed.

As I previously said, until you have visited a country, and no matter how much you read, nothing can beat the living experience.

Stepping out the airport that first time, I could not believe the oppressive humidity and heat, and later the terrible smell of sewage drains and stale cooking oil.

The traffic congestion and general pollution seemed to be part of the city makeup.

From early morning to late at night, the sound of jeepmeys could be heard no matter where you were.

The jeepmey came about when America left behind hundreds of their pickup trucks after the Pacific war ended.

The Filipinos soon modified these trucks to serve as public transport.

I have no idea what these drivers did to the thousands of jeepmey exhaust pipes, but I am sure they must have removed the silencers.

The jeepmey is used in most cities as the main way of getting around because it is the cheapest form of public transport.

These noisy vehicles can seat up to 16 passengers.

I guess they have become what the mini bus is too Africa.

No matter which country or city I visit, I have always made a point of travelling the way locals do.

Sitting in an air conditioned taxi or coach with the windows closed, you miss the sounds, smells and aromas.

Another first for me was travelling in a motorised tricycle.

These were used for shorter distances and only transported 2 people at a time.

No matter what time of day or night I landed and stepped outside the airport in Manila, it was always the heat and humidity that first hit one.

Another sound you will hear in Manila is street venders calling out One of the Filipino delicacies "balut."

This is a fertilised developing egg embryo boiled and sold by street venders as they walk from street to street calling out "balut."

Before you ask, no, I did not try it, the thought alone was bad enough.

Travelling by sound and smell

One other so-called street side delicacy were chicken feet. This reminded me of South Africa.

Talking about poultry, one of the things which saddened me about this country is how the government allowed cock fighting to take place.

There were even stadia set aside for this barbaric event.

Something else I could never get used to was the large portion of rice served with every meal.

Ordering breakfast at the hotel that first morning, I couldn't believe the large portion of rice which arrived on my plate with an egg along with a couple of slices of cucumber and tomatoes.

Even the bread was a no no. All bread is sweetened with sugar.

If you ever thought English food was bland, wait until you try most of the food in the Philippines.

Very few spices are ever used.

I am not a fan of fast-food chains, but if you have to resort to fast-food, and you don't want anything American, try Jollibee.

This Philippines based chain has opened branches in New York and London.

To give you an idea how popular Jollibee is in the Philippines, its profits last year were almost 4 times more than McDonalds reaching over $2.5BN.

The main difference is Jollibee sells a tomato based spaghetti with its chicken pieces. The spaghetti is very moreish.

On that first visit, I also could not believe how Americanised everything was. From the music heard on the radio, the number of American fast food outlets and coffee chains and public announcements all done with an American accent. It was like a mini America but without the guns.

I guess this love of all things American must go back to the introduction of Spam.

It was also unbelievable how many massive shopping malls there were in Manila with each trying to outdo the next in attractions and visitor experiences.

There was even a Venice Grand Canal mall. The waterway running through the mall is 200 metres long and 15 metres wide. The would-be gondolas are stationed underneath the replica Rialto bridge and Towering over the open plaza is a replica of St. Mark's Tower, also in Piazza San Marco.

Another mall which amused me was the 4 buildings making up the Green belt mall with a chapel in the middle surrounded by a small garden.

I wondered if the chapel served as a wedding venue, or somewhere where one could ask for forgiveness having previously given into temptation and allowed the devil to get the better of you as you spent your money?

It didn't take long to realise and appreciate the amount of poverty throughout the Philippines. It reminded me of South Africa, but in many ways was even worse.

The country is dominated by the catholic church. Filipinos are deeply religious and incredibly family focussed, and are also extremely friendly.

By the way, the catholic church has done nothing to call for the ban in cock fighting.

Mind you, I guess this church has other issues to worry about like excusing and denying rape and sexual abuse by many of its leaders throughout the world.

Like many countries in the region, if you married a Filipino, you married the family and probably the extended family as well.

It was sad to often see elderly Americans with a young Filipino woman at their side.

Travelling by sound and smell

While sitting outside enjoying a cup of coffee, I got to meet quite a few of these elderly gentlemen.

Many of these elderly Americans spent a few weeks each year in the country and took these young women to buy clothes and makeup instead of paying for their education.

Chapter 31

There are 3 things I remember about those flights to the Philippines with KLM on their B777 aircraft.

The first was probably the most strangest question I have ever been asked on any flight.

Between Taipei and Manila one of the passengers approached me and after some idol chat asked the following question "Do you sell ice-cream in Manila?"

I am not often left without words, but this question did this for a couple of seconds. It left me speechless.

Where this question came from I had no idea. I certainly was not wearing a shirt with ice cream images on it nor eating ice cream.

Once he returned to his seat, I began to wonder if I looked like an ice cream seller and what did an ice cream seller even look like?

The next thing I found strange was KLM offering its Manila passengers a South African liqueur called Marula cream.

When I asked the cabin crew why they were carrying this liqueur, they did not know, and in fact were unaware it was a South African speciality.

All I could think was the airline must have got bottles of this liqueur at a special discount.

One other thing I soon discovered was on all inbound and outbound flights to and from Manila no matter which airline, passengers were always offered fish and rice or chicken and rice. These were standard dishes in the Philippines.

On one of my return flights from Manila back to London on the

Travelling by sound and smell

KLM B777, I sat next to a man from the Netherlands.

He told me he was sick and tired of rice with every meal and was looking forward to a meal without rice.

At the time unbeknown to him and me every flight from and to the Philippines always included rice. Yes, true to form, chicken and rice or fish and rice was served.

I did smile to myself.

Chapter 32

During that first trip to Manila, I also boarded an Airbus A320 owned by Cebu Pacific a low-cost airline in the country and flew to Brunei to meet with some university students who were involved in conservation projects.

I have always been interested in conservation issues.

Back to that first flight of many with Cebu Pacific. All inflight fleet announcements are pre-recorded with what can only be described as a sickly sweet American accent. I could just imagine condensed milk dripping from the mouth of the person recording each announcement.

Brunei is one of the smallest and richest countries in the world. Its citizens pay no taxes and education and medical matters are free.

Sadly the students in Brunei were not prepared to speak out against the country's rulers.

If they did disclose what was happening, there would be severe financial consequences. Almost everything in Brunei is free to its 450000 citizens.

This was my first visit to an Islamic country and the difference between it and the Philippines couldn't have been more striking.

Landing in the capital Brandar Seri Begawan and walking into the airport, I could sense the change straight away. All the workers were men, and it felt incredibly oppressive. I doubt if the word smile was part of their vocabulary, let alone action.

This dour atmosphere continued no matter where I went.

Travelling by sound and smell

There was no laughter on the streets, and men and women were never seen together. It was a misogynist heaven not that they believed in heaven.

Tobacco products and alcohol are also banned in this oil rich country.

Thanks to its oil reserves it is one of the richest countries in the world.

Like many other Islamic countries all domestic workers and labour intensive work is done by mostly foreigners especially Filipinos.

I was not surprised to learn Brunei has the highest rate of obesity in Southeast Asia.

My Filipino guide in the country had worked in the country for 10 years and during that time had not once returned home to visit her 2 children.

There are more than 2 million Filipinos mainly women working throughout parts of Southeast Asia and the Arab world. Most are treated harshly by their employers.

In the most recent case in January 2023 a Filipino domestic worker in Kuwait was raped and murdered and her body dumped in the desert.

The government in the Philippines is also to blame for this situation as it makes millions from these workers each month who send money back home to support their families or their community.

More than 10% of foreign income is generated by these women sending back money to the Philippines.

I later discovered that most passengers boarding flights from the UAE, and other Islamic countries as well as Hong Kong, Singapore to and from the Philippines were Filipinos either returning home for 2 weeks or returning back to one of the countries where they were employed.

These routes were incredibly profitable for the airlines.

Most of the airlines going to and from Manila employed at least one Filipino as part of its cabin crew.

This individual would repeat the safety message in Tagalog, one of the main languages in the Philippines.

The Philippines airline Cebu Pacific was started in 1996, and mainly uses Airbus aircraft on its domestic and regional routes.

The mainstay of the airline is the Airbus A320. It also has the Airbus A321 and in 2019 and at the beginning of 2020 I managed to fly on their Airbus A320Neo and their recently acquired A321neo.

Chapter 33

There was one good thing came out of that first flight to the Philippines. I soon realised it would be possible for me to travel to Vietnam.

For years I had wanted to visit Vietnam. While in Manila I found there were several Filipinos who were prepared to go with me and be my guide.

The deal was I would pay them for their time and they would not only be my guide, but also take pictures for me in Vietnam.

When I was growing up, I remember hearing about that war but nothing about Agent Orange. it was only when I started reading Kindle e-books, I began to understand the injustice, the many American lies, the destruction of many Vietnamese villages and habitat and the deaths of millions of innocent Vietnamese civilians.

After arriving back in the Philippines once again on a KLM B777, the time had come to board a Cebu Pacific A320 flight from Manila to Saigon now Ho Chi Minh City (HcMc).

Something I could never understand about the American invasion of Vietnam were the millions of bombs dropped on south Vietnam compared to bombs on north Vietnam.

Don't forget it was south Vietnam America wanted to protect against the communist north. So why were 3 million tons of bombs rained down on the south while only 1 million tons on the north?

If the bombs were not bad enough, there was also more than 70 million litres of agent Orange sprayed over south Vietnam.

This deadly toxin caused many cancers and to this day babies are still born in Vietnam with numerous physical and mental deformities.

The greatest lie and injustice during that war was the use of Agent Orange by the American invaders.

This highly toxic herbicide was sprayed over vast tracks of jungle, rice fields, and alongside rivers. It contaminated the soil and waterways.

It didn't take long before the returning American troops began suffering from various cancers. For years the American government denied using Agent Orange.

You can read all about the American lies in my book VIETNAM 1964-1975.

Chapter 34

Visiting a foreign country for the first time makes no difference how many books you read, or the amount of online research you did. it didn't prepare you for the lived experience.

Unlike when my former partner Helen and I visited South Africa where I knew the people, knew the culture, knew where to go and knew all about the food, this was certainly not the case in Vietnam, Thailand or Laos.

In Saigon, there are probably about 4 million if not more scooters or mopeds. But again until you are there and experience the traffic, with the riders many times only inches apart from each other books and the internet are just hearsay.

All riders and passengers have to wear helmets in Vietnam.

It was during my first trip on the back of a scooter, I discovered how close to the next scooter I really was. I could stick my hand out and shake hands with my guide riding next to me as she clung to her rider in front of her.

Riding pillion or ze as they call it in Saigon allowed me to listen to the music blaring out from Karaoke bars, smell the aromas coming from food being cooked by street venders and listen to pedestrians talking and walking by.

Apart from the sounds of music and people talking the other sound which never seems to stop is the constant beeping of scooter horns. Fortunately these are quite high pitched and not that loud. The sound seemed to add a uniqueness to the city.

There are times when being blind is a definite advantage. You don't appreciate the traffic chaos and how many scooters are out on the streets.

Most first time visitors to Vietnam and in particular major cities are petrified at the thought of having to cross any street.

I know that almost all of the guides who went with me to Saigon were extremely worried at first when they wanted to cross a street.

They soon learnt there was a technique to crossing the road. If you waited for a break in the traffic, you could wait all day.

The trick was to ignore the almost continuous horn blowing and watch how locals crossed the road.

When ready, step off the pavement and cross at a slow, steady pace until reaching the other side.

Motorcyclists will avoid you as long as you don't step backwards, stop or make any unexpected movements. Always look both ways – even in a one-way street the traffic travels in both directions.

One thing I soon discovered when trying to walk on the sidewalks There seemed to be a lack of parking bays. Riders often parked their scooters on sidewalks as well as in the many alleyways.

Of course there are cars in Saigon, but even businessmen and women often use a scooter to get around the city.

I will never forget that first flight with Cebu Pacific on the A320 to Saigon.

It goes without saying when we landed late that night, there was a wheelchair waiting for me.

After clearing customs, getting a sim card and paying for the taxi fare before leaving the airport, my guide and I were driven to downtown Saigon.

Stepping out of the taxi on to the sidewalk, I was suddenly overwhelmed with emotions.

Some Popes might kiss the ground when they arrive in certain places, but I decided not to do this.

Travelling by sound and smell

I just stood there for a couple of minutes. I couldn't believe I was finally in Vietnam and it wasn't a dream. It was almost midnight and I experienced one of those déjà vu moments.

I had come home. Deep down I knew I had been there before. The next few chapters are devoted to Vietnam.

Chapter 35

Vietnam like many other countries is steeped in many legends and mythical creatures which are sacred animals representing the tremendous power of the universe. They are believed to be created from the four main elements including earth, water, fire, and air.

These mythical creatures can often be found in Vietnamese sculptures in pagodas, temples, cemeteries, and even local houses.

Making objects engraved with four mythical creatures often requires a certain amount of effort with high accuracy because the objects are usually used in sacred places and rituals.

Here are just a few examples, but during my travels in Vietnam, there appeared to be just as many local if not regional mythical legends.

One of the most important is the Vietnamese Dragon, a combination of a snake's body, lizard's thighs, hawk's claws, deer horns and fish scales.

It is considered to be the creature of heaven, possessing greater power than other animals and symbolizing nobility and immortality. People also think when the dragon appears, it will bring good things, luck, wealth and peace.

These are the reasons why the dragon was used as a special symbol of Vietnamese emperors in the past, and used to decorate the royal seal.

Vietnamese believe they are descendants of a powerful dragon named Lac Long Quan.

Travelling by sound and smell

Legend has it that Lac Long Quan married a beautiful fairy named Au Co. They had 100 sons and daughters, and the first-born son then became the king of the first dynasty of Vietnam: the Hung dynasty.

There is a Vietnamese proverb often used to describe themselves as descendants of dragon and fairy.

The features and roles of the dragon in Vietnamese culture are very different from those in Chinese.

The head of the Vietnamese dragon has a long mane, a chin beard, and no horns.

It also holds the pearl - a symbol of nobility, intellect, and humanity - in its mouth instead of in the front claws. Its curvy body represents its ability to change the weather and seasons, giving it an important role in Vietnamese lives and Vietnamese agriculture.

The next mythical creature is the unicorn, combining the features of a horse, dragon, and a buffalo.

Like the dragon, the appearance of the unicorn is widely believed to bring good fortune and peace.

You can often find the unicorn carved on doorways and unicorn statues in front of many pagodas and temples in Vietnam. It is because the unicorn has the power to guard houses, temples, and places of worship from bad spirits.

Legend has it that the Vietnamese unicorn was originally a wild beast living under the sea and coming to destroy crops and ruin normal life.

Maitreya Buddha, transformed himself into the Earth God in Vietnamese, domesticating the unicorn and turning it into a helpful beast.

This legend explains the origin of the Lion Dance which is often performed at traditional Vietnamese festivals including the Lunar

New Year and Mid-Autumn Festival.

In the four main mythical creatures in Vietnamese culture, the turtle is the only real animal.

As a long-lived amphibian reptile, the turtle has the ability to survive without any food for a long time representing longevity and spiritual strength.

The turtle also symbolizes immortal intelligence. At the Temple of Literature in Hanoi, there are 82 stone sculptures of turtles carrying the names of doctoral graduates carved on them.

This was seen as a mark of honour to those who had achieved the highest degree of education to serve the country during the Le Dynasty.

This creature is worshipped at the Ngoc Son Temple at the Hoan Kiem lake in Hanoi.

It is also closely related to Vietnamese culture through several historical tales. One of them is the tale in which Le Loi returned the sacred sword to the Dragon King, after he defeated the Chinese army, via a turtle that lived in a jade water lake.

The lake was later named Hoan Kiem lake which means "Returned Sword Lake".

The fourth creature is the phoenix, with its snake neck, eagle claws, fish scales, and a peacock tail.

This mythical bird symbolizes virtue, nobility, and grace.

According to legend the phoenix only appears in prosperous times, hence representing peace.

Originating from China, the phoenix is considered to be the most beautiful bird and one of the most powerful and sacred animals. Together with the dragon, the phoenix symbolizes royalty.

The dragon has a yang element, representing the king, while the phoenix has a yin element, representing the queen.

The combination of yin and yang symbolized by the dragon and

Travelling by sound and smell

phoenix is also believed to bring happiness to a marriage.

This is why the image of a dragon and a phoenix is often used as a decoration for Chinese and Vietnamese weddings.

Having learnt about some of the mythical animals of Vietnam, I couldn't help thinking about the South African living and widespread wildebeest.

The wildebeest is definitely not a mythical creature and must be one of the weirdest shaped animals in the world.

It has the head of a cow, the head and tail of a horse and the spindly legs of a gazelle.

Both males and females have horns which form curved semi-circles pointing slightly backwards. All have short hair covering their bodies with black vertical stripes of longer hair on their backs. Wildebeest have a black mane, which is thick and long. They have a long beard on their neck, which can be dark or pale.

Many people jokingly say God must have run out of parts when creating this antelope.

Apart from mythical creatures, Vietnam also has several UNESCO heritage sites.

As of 2021, there are eight World Heritage Sites in the country including five cultural sites, two natural sites, and one mixed.

The country holds the second-highest number of World Heritage Sites in Southeast Asia, after

Indonesia with nine sites.

Some of these sites include the monument complex at Hue which was the first site in Vietnam to be added to the list at the 17th session of the World Heritage Committee held in Colombia in 1993.

Built in the 11th century by the Ly dynasty, the Imperial Citadel contains buildings that parallel the late 19th-century architecture and the Southeast Asian culture. The site played an important role in the regional political power for almost thirteen centuries.

The site reveals royal power, new trends in technology and commerce in an imperial city. Its construction adapted the Confucian philosophy within a primarily Buddhist culture.

Another World heritage site was Hoi An, the ancient Town near Da Mang.

This ancient town with small shops located near the mouth of the Thu Bon River comprises timber frame buildings, which include architectural monuments, an open market, and a ferry quay. Its architecture reflects a blend of indigenous and foreign influences from Chinese, Japanese and European cultures. It is an example of a Southeast Asian trading port dating from the 15th to the 19th century.

The town is also known for its 400 year old covered Japanese wooden bridge.

Another heritage site near Hoi An is My Son, an ancient temple complex built between the 4th and 13th centuries, principally in dedication to the goddess Shiva, one of the foremost deities of Hinduism.

The temples in My Son have one of the most captivating stories and are one of the most interesting archaeological sites in Southeast Asia.

The history of My Son's origins through to its abandonment and enveloping lush jungle, along with the subsequent rediscovery of the ancient ruins centuries later by French archaeologists are fascinating.

The Champa Kingdom and its heritage are now largely forgotten, but for over 10 centuries it was the Cham people who owned and ruled the lands of contemporary central Vietnam.

Some theories say the Cham were descendants of Polynesian migrants who arrived on boats and chose these lands to settle

Travelling by sound and smell

down. They were without doubt, great sailors and at times even the pirates. They were some of the most important traders in the world, controlling one of the main trade routes in the region.

Sadly what is left of their former greatness are mainly ruins and old temples in several parts of Vietnam.

The most magnificent of them is the My Son Sanctuary.

Both My Son sanctuary and Hoi An old town are close to each other. They represent the history of the lands of contemporary Vietnam at a time when the Dai Viet people were still under Chinese occupation and were fighting for their independence.

The sanctuary represents the abiding legacy of the once prominent Cham culture that now barely exists but was at one point as strong and large as the Angkor Empire, famed for building Angkor Wat and other temples in Cambodia.

In fact, the ancient Champa and Angkor Kingdoms were neighbours and rivals but shared their culture and beliefs. The Cham, just like the Khmer, were mostly Hindu.

They adopted Hinduism directly from the Indians, who they were extensively trading with.

The My Son Sanctuary was built over the course of 900 years and is an insight into the history, religion, and politics of that time. It is also a wonderful display of the cultural sophistication of the kingdom and the fascinating artistic skills of its people.

When the Dai Viet people from the north freed themselves from the 1000 years of Chinese occupation, they started to march southwards, away from the enemy and towards the expansion of their own country. War after war, eventually the Cham had to abandon their sacred land and move south, finding new settlements.

The youngest My Son temple dates back to the 13th century.

After that, the jungle took over, engulfing the temples until the French scholar, M. C. Paris, with the help of local farmers in 1898 found the ruins.

The French School of the Far East began research on the site shortly after the discovery.

The word temple implies a large structure, but the Cham kings built rather small, red-brick structures to house the lingam and yoni representations of Shiva as well at some stage statues of Buddha.

Not all buildings at the complex served as shrines, some were treasuries, others as baths and changing rooms to prepare for the service.

The temples represent the sacred Mount Meru, the centre of the Hindu universe. Just as important as the structures themselves are the sandstone and brick reliefs, depicting deities, guardians, and events.

Many structures have been lost over the centuries, but the remaining ones are truly impressive. A big collection of Cham sculptures is on exhibition at the museum in Da Nang.

The restorations that started in the 1930s brought back a lot of its former splendour but the US bombs and landmines destroyed much of the sites.

Thankfully some 70 temples remain fully or partly intact and even some intricate decorations have been preserved. The site is constantly under renovation, bringing more and more temples to their former glory.

In recent years the pathways and infrastructure have been hugely improved.

If you visit the area, you will be able to walk through the sanctuary and see the temples and ancient function buildings in several groups. You can enter many of them, and see the art up close.

Visitors will also notice the war damage, including many bomb

Travelling by sound and smell

craters that have been purposely kept at the site as an important part of history.

Ha Long Bay is another UNESCO site featuring more than 1600 karst limestone pillars and isles in various shapes and sizes, developed in a warm and wet tropical climate.

The limestone monolithic islands rise from the ocean, topped with thick jungle vegetation. Several of the islands are hollow, creating enormous caves.

In addition to its World Heritage Sites, Vietnam also maintains seven properties on its tentative list. This list is necessary for future sites being declared as UNESCO World Heritage sites.

Chapter 36

Prior to travelling to Saigon, I found and made contact with a university student group calling themselves Saigon Free Walking Tours.

This group of students offered to take tourists on the back of their scooters to various attractions in the city.

The idea behind the programme was to allow those students to practise their English and also promote the various attractions in Saigon.

There was one walking tour while the other 7 attractions involved going to these on the back of a scooter.

During the following visits, I got to meet several students, including Tuyen, James, Julia and Linh, Jinny and Danny to mention just some.

Once again, you can read all about that first visit in my book (Guiding Hands and Wheels).

Apart from making prior contact with some of the students, I had also chosen a hotel to stay at.

My choice was simply based on its name. It was called Pink Tulip Hotel, which turned out to be a great choice.

In the trips that followed, I got to know the owners, the staff and much about their families.

After that first western breakfast, I asked the receptionist Thao if I could rather have what locals ate.

The next morning she arrived on her scooter with a bowl of pho noodles. This dish is found throughout Vietnam.

Travelling by sound and smell

In the next few weeks I got to try many Vietnamese foods as well as enjoying many thirst-quenching freshly squeezed fruit drinks.

After my bowl of pho the next that morning was freshly squeezed sugar cane juice when we went out with our guides Tuyen and James from Free Walking Tours.

If you read on, I will give you some of my favourite recipes.

Chapter 37

One of the first things I fell in love with after landing was the tonal language. It sounds like singing with up and down accents. There are 6 different tones used in the language.

Believe it or not, there are more than 4 million Vietnamese speakers outside the country with most found in the Czech Republic, America and Germany. In 2019 Vietnamese was ranked 17th for most spoken native languages in the world.

For most of its time Vietnamese was an oral language passed down by generations.

Vietnamese culture, food and language were heavily influenced by the Chinese and French invaders and colonisers. This is also reflected in many of the foods often with a Vietnamese twist to them.

Thank God none of the American food chains have caught on in the country.

Vietnamese is an extremely difficult language to learn.

For example, there are 6 different ways to address a person.

Ways to address a person changes depending on age, family or social status and even the relationship towards the person.

A person slightly older than you can be addressed in one way depending on the gender.

If a person is your parents' age, but younger than them then another form is used, but if, older than your parents then yet another form is used.

If the person is significantly older and depending on their social status, then another form is used.

It is terribly confusing, but I still recommend you learn a few words before visiting the country.

Another thing I soon found out was most young people when dealing with visitors use an English name instead of their birth name.

Something else I learnt and got to admire was how humble the Vietnamese are and how unlike the West, the community comes first and not the individual.

But more about this humility and attitude later.

We could all learn a lot from this society and its people.

Chapter 38

After Tuyen and James from Saigon Free Walking tours arrived and following introductions, we walked downtown with them describing some of the well-known sites.

Fortunately that morning it was reasonably cool, or certainly for Saigon weather.

We spent a bit of time inside the famous post office, and then did a tour of the nearby Reunification Palace where the North Vietnamese army tank crashed through the iron gates on April 30 1975, bringing an end to the Vietnam War.

Set on 44 acres of lush lawns and gardens, the palace also offers a fascinating glimpse at the lifestyle of privileged heads of state in 1960s Saigon. It was built on the site of the former Norodom Palace, which was bombed by fighter jets in 1962 in an unsuccessful assassination attempt on the South Vietnamese president Ngo Dinh Diem.

The current building was completed in 1966 and became the home and workplace of the then president, when Vietnam was split between the north and the south.

Some of the features include the president's living quarters, the war command room with large maps and antiquated communications equipment, and the maze of basement tunnels.

There are also military vehicles outside, including the fighter jet that destroyed the Norodom Palace, and tank 843, which rammed through the palace gates on that fateful day.

I was able to touch and feel the tank and plane.

Travelling by sound and smell

Fortunately the old opera house, the War Remnants Museum and the Notre Dame Cathedral were all nearby.

Tuyen and James told me that the outside red bricks of the cathedral were imported from Marseille, and the clock between the two bell towers was built in Switzerland in 1887.

We finished the morning walk by visiting the famous Ben Thanh market with its thousands of traders.

On our way to the market, our guides asked me if I knew about some of the drinks sold on the streets of Saigon.

At that stage, no, I had no idea.

Before we entered the market, James found a vender selling squeezed sugar cane juice for me to try.

This is when I got to try the freshly squeezed sugar cane juice. They also told me about sugar cane juice sometimes with added ginger, as well as coconut water with kumquat, smoothies made from a variety of blended fruits, lotus seed milk, sweetcorn milk, mung bean milk, peanut milk, and various other fruit juices.

What I didn't realise, unlike fruit juices back in the UK, these were not in bottles or cartons, but were made while you wait.

They assured me there was even more to look forward to when it came to coffee.

It wasn't long before I discovered that most locals did not bother about eating at restaurants, but ate from the many street food venders.

They also suggested I should try the famous banh mi, a version of the French baguette.

The market is like a one-time shop offering everything from fresh fish, vegetables, meat, spices, gifts, clothes, kitchenware and so much more. There were also lots of fruit and vegetables I had never heard of or tasted.

While walking through the market, I decided Vietnam must be

the fruit heaven of the world. I can't think of a single fruit not grown in this beautiful country.

During the next hour Tuyen described most of the fruit to me, some of which I was able to touch.

As we made our way past the numerous fruit stalls, she also told me about the health benefits of each.

Those I did not know included Rambutan with its hairy skin on the outside.

Apparently the typical colour is yellow, greenish-yellow, and red when ripe.

It is a rich source of fibre and vitamin C, copper, manganese, potassium, phosphorus, magnesium, iron, and zinc.

Longan is a round fruit with leathery yellow or brown skin. Its name means 'dragon eyes' in Cantonese. The reason behind its name is the black seed covered in translucent white flesh looks like dragon eyes.

This summer fruit contains carbohydrates, fibre, vitamin B1, vitamin C, but has a high sugar level.

Ambarella is oval shaped, green or greenish-yellow, with a spotted skin. The flesh is white or slightly yellow and crunchy.

The closer the flesh to the pit, the sweeter it is.

Tuyen said most Vietnamese people believe Ambarella helps with the recovery from colds.

It contains fibre, carbohydrate, potassium, phosphorus, and vitamin C.

Sapodilla was brought to Vietnam from Thailand, and has thin, light yellow to brown skin. It usually has several black seeds. The flesh of Sapodilla is very soft and sweet when ripe.

It contains fibre, tannin, polyphenolic, calcium, vitamin A, and antioxidants.

Remember to remove the seeds before you eat them.

Travelling by sound and smell

Lanzone or Langsat is native to Southeast Asia. In Vietnam, it is grown throughout the country, and is found in markets and supermarkets. This fruit is round with a soft rind, translucent white pulp, and five to six segments.

Green lanzone is sour but gets sweeter and changes to yellow when ripe.

Tuyen told me people must avoid eating its rind and seeds since they contain a small amount of toxicity. She also said people who need to watch their sugar level should not eat it due to its high level of sugar.

It contains Antioxidants, fibre, vitamin C, vitamin B, and vitamin E.

Jackfruit is one of the most fragrant and sweet fruits in Vietnam.

The fruit originated in India and is mostly grown in Southeast Asia and Brazil. In Vietnam, jackfruit is a popular fruit and can be found nationwide.

Jackfruit is a big, if not giant, fruit with bumpy and rough skin. However, the flesh inside tastes delicious. She said jackfruit is very fragrant, and when it is placed inside a house, everyone will know it is there.

It contains Vitamin A, vitamin C, calcium, potassium, and iron.

Mangosteens are also very popular in Vietnam.

This fruit is round in shape and dark purple in colour. It has a stem with 3 to 4 leaves on one side and a small flower on the other.

People say that the flower is used to show how many segments are inside.

The white flesh of Mangosteen is divided into segments with or without seeds inside.

She said this fruit tastes sweet and slightly sour. It is also a very healthy fruit containing Antioxidants, potassium, carbohydrates, fibre, folate, and manganese.

Another popular fruit in Vietnam is dragon fruit, and is red and round with many spikes on the skin.

It has many edible black seeds inside the flesh, and removing them is nearly impossible.

I never knew it belonged to the cactus family.

There are two types of dragon fruit in Vietnam: white flesh and red flesh.

The fruit tastes mildly sweet and slightly sour. People say that they taste best when chilled.

Tuyen said in Vietnam a special type of bread is sometimes made from the red flesh of dragon fruit. With its special pink colour and flavour, Dragon fruit bread is very popular in the country.

This fruit contains Antioxidants, vitamin B2, vitamin B3, vitamin C, calcium, potassium, and phosphorus.

Because of its spikes, Tuyen warned me not to touch the fruit.

In my many visits to Vietnam I got to try all the above with the exception of Durian.

This fruit is apparently a tasty and nutrient-rich fruit despite its strong smell.

It is a tropical fruit native to Southeast Asia, and looks similar to Jackfruit, albeit with sharper and longer spines.

Due to its strong smell, Durian is prohibited in some restaurants and on public transport.

It can be eaten raw or turned into a sweet dessert.

While it apparently has a foul smell, it is rich in Carbohydrates, vitamin B6, vitamin C, thiamine, potassium, manganese, folate, copper, and magnesium.

I doubt if there is another country which grows so many different varieties of fruit.

Other fruit apart from those already mentioned include grapes, avocado, plum, strawberry, mulberry, coconut, banana, passion

Travelling by sound and smell

fruit, pomegranate, kumquat, pineapple, papaya, red and white guava, pomelo, oranges, mandarins, star fruit, various types of apples, sugar cane and these days seedless watermelon.

As we walked back to the hotel Tuyen and James told me how fruit plays an important role during the Tet celebrations.

She said watermelon is one of the favoured fruits during the Tet holidays. The local people carve lucky words and pictures on the skin to pray for happiness and prosperity.

I asked her how come she knew so much about each of the fruit on offer.

She told me as someone interested in all things food and fruit the more one learnt about the subject the better informed one was.

I guess we could all learn from her words.

Ben Thanh is mainly a tourist attraction even though residents also shop at the market, but there are numerous smaller neighbourhood markets where the locals buy and barter.

At night scores of pop-up food stalls around and in the market open up.

During that first visit I also went to the Cu Chi Tunnels about 60 kilometres from Saigon. The tunnels are a vast 250-kilometer-plus underground network used by the so-called Viet Con.

The tunnel networks often included underground hospitals, workshops and some entertainment areas.

Believe it or not, the Americans unknowingly built a base above this network.

I was able to touch the various booby traps, and also sat at the entrance of one of the tunnels.

Feeling the various booby traps it reminded me how inhumane people can be.

You can also sit down and have a traditional meal which the North Vietnamese soldiers ate during the war, and also buy a pair of uncle Hoe sandals.

We also spent a day on the Mekong river visiting a couple of islands and sampling local food.

Before that first trip, I had decided I also wanted to visit the many different coffee shops. Unlike various themed bars in the West, many Vietnamese coffee shops instead offered themed coffee shops.

This was also my first time to taste iced coffee. That was it, I was hooked and could never get enough.

Vietnam is famous for its robusta coffee beans.

For me the Robusta bean is far superior to the better known Arabica bean.

The next 2 days were spent with Tuyen and James taking my guide and I to numerous coffee shops.

In the following trips to Vietnam on one of the Cebu Pacific A320 aircraft I got to meet and go with several more students who were part of Saigon Free Walking Tours.

One of the things I soon learnt about the Vietnamese people was their humility. Also in all my visits I never saw a drunk and aggressive local. It was always those tourist from the West, who would be seen out on the streets drunk and shouting.

Buddhism plays an important part in Vietnamese life and attitude. Much to my surprise there was hardly any recrimination and even discrimination against American tourists.

I think this attitude had to do with their philosophy to live in the present and not dwell on the past. We could all learn from them.

On one of my later visits, I later discovered how Abba and their song (happy New Year) played an important role in Vietnam festivities.

More about that song and the festivities later.

Chapter 39

My trips from the Philippines to Vietnam and Thailand were always on one of the Cebu Pacific A320 aircraft.

Most times the young cabin crew on these flights failed on take-off to put my white cane in the overhead lockers.

Cebu Pacific seemed to employ only young women as cabin crew.

That first trip to Vietnam also found me visiting several temples and pagodas. However, when you are not allowed to touch things, temples and pagodas were not going to feature on my got to do list.

After a couple of temples and pagodas, I was so to speak templed out.

Some of the following flights to the Philippines after those with KLM on their B777 aircraft was with Emirates on their A380 to Dubai and then on to Manila on the B777.

Other airlines which took me to Manila included Turkish, Kuwait, Qatar, Thai Airways and Cathay Pacific.

Previously, I had travelled with Helen on the A380to South Africa and immediately fell in love with this aircraft.

To say I had become an A380 fan would be an understatement.

Mind you, any Airbus will do.

After flying on numerous Boeing aircraft including the B720, the B727, the B737, the B707, the B757, the B767, various B747 aircraft from the B747-100 through to the B747-400, and the B777 were all noisy compared to all Airbus planes.

The only Boeing aircraft which was not noisy, was the B787 known as the Dreamliner.

John CT Miller

Try as I may to only fly on Airbus planes I couldn't avoid Boeing.

In January 2023 after 54 years the production of (the queen of the skies) or as most people called it (the jumbo jet) as it was also known, the B747 and its numerous variations came to an end.

Those 5 decades Boeing sold 1574 B747 aircraft. The final a 747 freighter version was delivered to Atlas Air.

The Boeing jumbo in its time carried over 4 billion passengers around the world and several space shuttles.

Sadly 64 B747 jetliners have been lost in accidents and incidents and 3,746 people have died.

I know the number of deaths might sound a lot, but remember, air travel is by far safer than riding a bike, driving a car and safer than crossing a street. deaths might sound a lot, but remember, Previously for every 350000 passengers there was one death.

In the 90s it was one death to every 1.3 million passengers and today it is one death to about 8 million passengers.

When it comes to flying on Airbus aircraft, the list includes the A19, A20, A21, A320, A330, A300, A340, A340-300, A340-600, A380 and their A350 aircraft.

Chapter 40

While on one of my visits to Vietnam with a dear friend Jacquilen we met an Australian couple at Pink Tulip hotel, who suggested we should take a flight to Da Nang and visit Marble Mountain and also the nearby town of Hoi An.

Sadly it was towards the end of our visit to Saigon so it would have to be a flying visit so to speak.

Da Nang and Hoi An would find me back there later with another friend Donnah.

That first trip to Da Nang was also my first flight with the local budget airline VietJet.

VietJet previously known as the 'bikini airline' is one of the most successful low-cost airlines in Southeast Asia.

Since launching in 2011, the airline has published an annual calendar that features women in bikinis posing as cabin crew members such as flight attendants, pilots and ground staff.

The airline is the Brain-child of a woman entrepreneur and founded by Vietnam's first self-made woman billionaire Nguyen Thi Phuong Thao.

In 2012, the airline found themselves in trouble when five female bikini-clad flight attendants took part in a mid-flight dance posing as beauty pageant contestants. The airline had not obtained permission from Vietnam's aviation authorities, and was fined £678.20.

Since then it has become a favourite with the travelling public.

The airline started with several Airbus A320 aircraft.

I was lucky enough to fly to Da Nang on both the Airbus A321 and the Airbus A321neo.

That first trip to Da Nang we visited Marble Mountain and after that went to Hoi An.

Hoi An along with Da Lat has become 2 of my favourite places.

Each night the UNESCO site Hoi An is lit up by hundreds of lanterns found in trees and outside shops and houses.

Hoi An is probably best known not just for its wooden buildings, but also for its number of tailors.

Just by the way, it is also the place where you will find the world's best black forest cake.

Jacquilen and I accidentally found this restaurant opposite the river while walking down the street.

Not only did it bake the best black forest cake in the world, but this hidden gem also served a delicious and decadent chocolate and passionfruit cake.

For those looking for picture opportunities, you have to visit Da Nang's illuminated bridges.

If you are in the city over a weekend the most famous bridge the Dragon Bridge is a must. This bridge not only slowly changes colours, but at weekends the Dragon bridge spits out a wall of fire followed by a wall of water spraying from its mouth.

One other not to miss venue in the city is the Thanh Tam Special School which looks after and cares for children suffering from Agent Orange.

The school has its own coffee shop and organic bakery on the premises.

It really is simply the best organic coffee and bakery in Da Nang. Apart from their selection of homemade cakes the shop also serves breakfast and lunch, and better still employs a couple of disabled people in the shop.

Free internet is also available.

Travelling by sound and smell

Patrons are also encouraged to spend time visiting the school and their arts and craft shop.

Just like the black forest restaurant in Hoi An, we found Thanh Tam school and bakery by accident.

This became my favourite eatery, not just for its food, but unlike commercial restaurants the money they made went towards helping the school.

While Vietjet might be known as the "bikini" airline, for me, it became the Abba airline.

Let me explain.

On one of my trips I happen to spend new year in Saigon and Da Nang and on another, I spent part of the annual Tet celebrations in the city.

On each occasion, not only did you hear Abba's "Happy New Year" being played by live bands and solo artists in many Saigon streets, but also heard the song playing in many shops.

If you thought you had heard enough of Abba, you were in for a surprise when boarding one of the VietJet airbus aircraft the song was on a loop and continued playing until take off.

As soon as we landed in Da Nang and on entering the airport terminal "Happy New Year" was playing again.

Da Nang airport was once one of the busiest airports in the world during the American war in the 60s.

It took over 30 years before the Agent Orange contaminated grounds near the airport were made safe.

Da Nang's military museum is worth a visit. It features 3 thousand years of military history, ancient weapons and those used and captured during the American invasion of Vietnam.

The city is also not that far from hue, the former royal capital, and once again the food on offer is different to that found in other cities.

Chapter 41

Sadly the owners of Pink Tulip in Saigon sold the hotel and decided to move to Hoi An.

But before they moved, they introduced me to the owners of Saigon Amigo hotel which was almost opposite Pink Tulip.

During my following visits, I got to know Li the owners daughter who ran the hotel and met her parents and other members of her family.

Once again, you can read all about my trips to Vietnam in my last book titled (Remembering Past travels) and thanks to Li, got to hear 2 of the most beautiful Vietnamese songs.

I will always remember Saigon Amigo Hotel for hearing 2 of the most beautiful songs I have ever heard.

While drinking my 3rd of 4th iced coffee, suddenly this song began to play on Li's sterio.

I even stopped sipping my coffee to listen to it.

When it finished playing, I asked her to play it again.

The 2 songs are called: EM NGAY XUA KHAC ROI

Tying vi-duong Hoang yen, and Giau-trang phap, both by the singer HIEN HO

Check them out on YouTube.

Before I give you a few of the most popular Vietnamese dishes, I really encourage you to first go on a street food tour.

Travelling from the airport into Saigon, you can't help but notice the many child-like plastic chairs and tables set out across the city. Many are only a few metres from busy roads.

Travelling by sound and smell

Vietnamese families and friends can often be found sitting around these tables enjoying the various foods on offer.

It didn't take me long to realise the variety of street food was the same as that on offer at restaurants, but a lot cheaper.

One other reason for choosing street venders: the money earned goes directly to them.

Street food is available from early morning until late at night.

You will also see many venders on scooters going up and down streets selling various foods.

Before giving you a few recipes, for those lovers of ice-cream, you must try the coconut ice-cream.

Other ice-cream blends to try are avocado and macha.

While on the subject of yummy desserts, a definite must to try is called sua chua nep cam, a purple sticky rice yogurt.

Sitting on the banks of the Saigon river and chatting to 2 of the students from Saigon Free Walking tours Jinny and Danny I got to try this delicious dessert.

As we enjoyed the dessert, Jinny told me about her days at school and her belief in wanting to help others.

She told me her secondary school was opposite the war remnant museum, and was able to visit it for free.

She said she was able to witness some aspects about the Vietnam war, including the effects of agent orange and its victims.

"My parents were born before the war stopped 4-5 years. They told me that although it was coming to an end, and they were still little kids, they could still feel its brutality. When there was an agent orange victims topic on TV, my parents can only utter a cry of pain what the victims have been through.

They also taught me that my generation is very lucky to live in such peaceful environment like this, and that we should cherish, maintain and better the world by being good people and doing nice things. Because war only brings separation, pain, and death."

She told me she had taken part in different community services when she was at university. In March, 2017, she became a volunteer in Hoa Sen University's Art therapy intervention project for orphans at SOS Village.

In August that year, she applied again for the same project for street children at Thao Dan Social Protection Centre.

Jinny said "Throughout these projects, I have learned how to become a productive planner and an efficient conversation holder with the teenagers and children. This was where I found out that listening skill and empathy are also my most valuable possessions.

I continued getting involved in English teaching project for children at Thao Dan Social Protection centre.

I decided to do volunteering because my parents told me to "cherish, maintain and better the world by being good people and doing nice things". Moreover, unlike other students would choose to work part time to earn their livings, I chose to do community service. First of all, because I'm luckier than them that I have a scholarship, family support and sometimes, got tips from Saigon Free Walking Tours. But that's not the main reason, even if I don't have scholarship or else. I just don't want to involve in money, but rather build my knowledge and experience instead.

I'm kind of an ambitious person, I don't know if I use that word right, but I have the thought that when I do something, I want it not only to be good for just me, but the society have to receive something also, or at least for the community that I serve.

I want to have a positive influence on other people and together, make the world become better. I just want to step by step, build a better world with countless good people out there."

Listening to Jinny, I just thought if only we all followed her example, what a better world it would be.

Travelling by sound and smell

I know that Danny had also done some charity work, but not sure where and for which charities.

She and Danny were typical examples of the Vietnamese belief system in putting the community first.

Now let's get back to all things food.

Black sticky rice has some other names such as black glutinous rice and black sweet rice. It is a long-grain rice with uneven colours. Overall, it has a dark purple colour, and you can see that very clearly after soaking or cooking it. The contrast between its dark purple colour and the creamy white colour of yogurt creates such a striking visual effect.

Another interesting Vietnamese recipe that uses yogurt is Yogurt Coffee (Sua Chua Cafe).

Sticky rice has a nutty and rich flavour, and its fragrance is also stronger than regular sticky rice.

Some articles say that black sticky rice is a nutritious grain. It is rich in protein, antioxidants, dietary fibre, and iron. It may have anti-inflammatory effects and benefits for heart health.

Believe me, once you have tasted it, you will be back for more.

How to make this pudding?

First, wash the black sticky rice and soak for 4-6 hours. If you soak it for a shorter amount of time or skip this step, you will need to increase the cooking time.

After soaking the rice, you just simply simmer it with water until the rice is soft and the rice pudding is glossy and thick to your liking. Simmer the rice with pandan leaves to add some extra aroma, but if you can't find them simply add some vanilla extract and cook for 20 minutes.

Once the black sticky rice pudding is ready, Transfer to a bowl and let it cool to room temperature.

While you are waiting for it to cool down, whisk the plain yogurt until smooth before mixing it with the sticky rice.

Put 1 part black sticky rice pudding and 2 parts yogurt in serving glasses or bowls. The ratio of rice pudding and yogurt is up to you.

Add crushed ice and honey if desired and get ready for a second helping.

Two of the best known foods are Banh Mi, a Vietnamese styled baguette, and Pho noodle soup.

These sellers are everywhere.

Originating in Saigon, Banh Mi is a Vietnamese sandwich made with crispy, airy baguette typically filled with savoury beef, pork or chicken, mayo, a smear of rich liver pate, topped with pickled veggies, crunchy cucumbers, carrots, spicy chilies, fresh herbs and Thai basil.

Banh Mi is a fusion of Vietnamese and French cuisine brought about in the mid- 1800s, by the French occupiers and invaders in Vietnam.

You can see both influences clearly. The French styled bread, but with the pate and mayo,- and the Vietnamese, who lightened up the French baguette with rice flour, and added all the fresh and spicy ingredients.

I am not going to offer you a banh mi recipe. I am sure you will find a few on the internet, but believe me once you have taken a bite of your banh mi after watching the street vender put together your banh mi, you will be back for another.

You won't have to go far to find someone else selling or driving past selling banh mi. If you don't see a motorised seller you will definitely hear them as they use a pre-recorded amplified advert.

The other dish which is found all over Saigon and most of Vietnam is pho noodles.

Pho properly made is a deeply spiced and aromatic soup with

a clear broth with rice noodles, topped with different cuts and textures of meats and fresh herbs.

It doesn't need to be meat, and in fact many Vietnamese have vegan pho, but it's by default beef noodle soup, which is what this recipe is for.

A great pho soup is a well-made beef broth with charred aromatics, sweetened with sugar, and seasoned with fish sauce. The signature taste of beef pho though comes from the charred aromatics and spices. If you leave those out, you don't have pho, but beef noodle soup.

Charred herbs and vegetables are the most important part of a great pho soup.

Every good noodle soup needs fat for the broth to stick to the noodle and pho is no exception.

Leave the pho in the fridge until the fat solidifies (typically 4-6 hours) and then Scoop only the fat out into a small saucepan.

Melt it over low heat, and strain into a small container.

The fat will keep in the fridge for up to a week, and you can measure out exactly how much each bowl will have. Add at least a teaspoon to each bowl of pho you make.

Pho is about balance, but not necessarily subtlety.

You will also need sugar and fish sauce.

If you can, get fresh pho noodles, but if you can't, the dried stuff works too. Sometimes the noodles will be called rice stick or Thai rice stick noodles. Medium thickness is best.

Briefly blanch the noodles about halfway to your desired softness, then drain and rinse in cold water so they don't stick too much. Allow them to dry out in a colander for 5-10 minutes while you prepare the other items.

Drying out the rice noodles seems counterintuitive since you just cooked them, but it's the secret to flavourful noodles as they suck

John CT Miller

in the pho broth to rehydrate later.

Pho is loved not just for its broth but the meaty toppings. Sirloin, well done brisket, flank, tripe, meatballs, the list goes on and on.

A good pho should have 2-3 different meats. If you make this with finger meat, flank, or brisket, you should have some nice well done meat, and to that I'd recommend some meatballs, sirloin that's been thinly sliced and allowed to cook in the broth.

Vietnamese foods are known for their fresh flavours and herbs and pho is no different.

Sliced raw onions and chopped cilantro and Thai basil are a must, while Bean sprouts add an earthiness Sriracha and hoisin sauce

Top with onion and cilantro, then at the table, let everyone add Thai basil, bean sprouts, chilies, hoisin, and sriracha to their liking.

How to make pho on the stove

Char your spices and vegetables in a broiler or cast iron pan, just char the onion/shallots, ginger, and cinnamon.

Use beef bones to make a broth then drain. Ideally, wash the pot again before using.

Fill up the pot with the parboiled bones, aromatics, and spices along with enough water to cover. Bring to a very gentle simmer and leave it there for 4-6 hours. If you have flank or brisket, add them in 2 hours before the broth is finished cooking.

Just before you intend to eat, prepare the rice noodles,

Travelling by sound and smell

meats, and garnishes while the broth is cooking.

Add sugar and fish sauce, or salt until the broth is just about over seasoned. It'll balance out once you have added the noodles.

Place the rice noodles and sirloin in your bowl, then fill the bowl up with about 2 cups (or more) of pho soup. Top with onion and cilantro, then at the table, let everyone add Thai basil, bean sprouts, chilies, hoisin, and sriracha to their liking.

One of the secrets to pho, is to serve it in a deep and large bowl and gently heat the bowls before you add the ingredients.
Another favourite is Vietnamese Broken Rice with Grilled Pork (Com Tam), which was once known as "poor man's rice."
Broken rice wasn't always a popular fast-food option. Eating it was once a necessity that came with a stigma attached.
Poor rice farmers in the Mekong Delta started eating the broken grains because they could not sell them — and they soon developed a taste for them.
Urbanization in the first half of the 20th century brought the dish to Saigon where it quickly became a working-class staple. Grave food shortages in the 1980s, brought about by the failings of collective agriculture, shattered whatever class divides still lingered.
After the American war, Vietnam soon became one of the largest exporters of rice, but they could not sell the damaged rice.
What was once reserved for the poorest of the poor eventually became a "standardized part of the culture."
Soft and succulent, but never mushy. broken rice is somewhere between sticky rice and regular rice. Some even liken it to couscous.

John CT Miller

Apart from being cheaper than regular rice, it also cooks faster, which means lower energy bills.

Cooked properly — broken rice must be soaked before cooking or it becomes a starchy mess.

Ingredients

2 boneless pork chops, 1 tbsp oil, 2-3 cloves garlic (minced), 1 tbsp soy sauce, 1 tbsp honey and 1 tbsp fish sauce.

For the marinade you will need 1 tbsp vinegar, 1 lemon (juiced), 3 tbsp fish sauce, 1 to 2 tbsp sugar, 3 cloves garlic (minced), 1 to 2 cups warm water, 1 to 2 tbsp shallot (minced), 2 eggs, 100g minced pork, 1 tbsp oil, 1 tbsp fish sauce, ½ tsp salt and ½ tsp pepper.

For Serving

2 cups of broken rice (cooked), 2 fried eggs, Pickled carrots and radishes, Cucumber (seeded and sliced), Soy sauce (to garnish), Spring onion (chopped and sautéed slightly).

1 Place all the marinade ingredients in a bowl and mix well until sugar is dissolved.

To Cook the pork chops, mix the remaining ingredients until well-combined. Coat the pork chops with the sauce and marinade for at least 1 hour or overnight.

Travelling by sound and smell

3 Heat a grill or frying pan with oil. Grill the meat until done, about 1-2 mins on each side. Remove and set aside.

To Make the Meatloaf, set 2 egg yolks aside, add the remaining eggs into a bowl and beat well, and

add the rest of the meatloaf ingredients and mix well.

Then Grease a square or loaf tin and place the mixture into the tin. Spread evenly and cover it and steam for 30 mins.

Spread the saved egg yolks on the top of the meatloaf and steam uncovered for 5 min. Remove and allow to cool before slicing.

To Serve, in a serving plate, place the broken rice, grilled pork chop, meatloaf slice, fried egg, cucumber slices, pickled carrot and radish slices. Garnish with sautéed spring onion and serve with dipping sauce.

Remember each city or region has its own special dishes, but Pho and banh mi can be found throughout the country.
    When I was in Da Lat, I managed to find someone at the local night market making a Vietnamese pizza.
    This is nothing like a pizza found in the West, but even better.

Chapter 41

Another country I visited on both the A320 and the B777 was the land of smiles or Thailand as it is known.

I think it should also be known as the land of shrines, Buddha's and monks.

The suvarnabhumiinternation airport in Bangkok has a unique feature not found in any of the other airports I visited.

After clearing customs, and before you leave the terminal you will find at street level a mock-up of a Thai street market.

This is a must place to visit after your flight and before you leave for the city or catch a bus to other parts of the country.

At this indoor airport market, you can find something to drink and begin to get to know some of the many Thai dishes on offer.

Much like the Vietnamese, the Thai people are extremely humble and thankfully the language is a lot easier to learn.

Bangkok with its population of 10.72 million permanent residents as of 2022 just like Manila and Saigon is a highly congested city.

But, that's where the similarity ends.

Unlike the other 2 cities it does offer visitors and locals many different ways of getting around the capital.

With trains, buses, subways, boats, sky trains, taxis, tuk-tuks, and songthaews all waiting to transport you, there are so many options.

When navigating your way through Bangkok's public transportation system, it's important to be clued up on the route maps and how they work, so you can get from A to B without a hitch.

Travelling by sound and smell

The songthaews are converted pick-up trucks, with two bench seats fixed along either side of the back of the truck. Some vehicles have roofs high enough to accommodate standing passengers within the vehicle.

Songthaews are used both within towns and cities and for longer routes between towns and villages.

Those within towns are converted from pick-up trucks and usually travel fixed routes for a set fare, while vehicles on longer routes may be converted from larger trucks carrying about forty passengers.

As you already know, I prefer using public transport, which is the best way to meet locals and truly experience local life.

They reminded me of the jeepmey used in the Philippines.

A word of advice. If at all possible, do not use a taxi. Because of the traffic, taking a taxi is slow and will cost a lot more compared to the various public transport choices.

The transport options start the minute you leave the airport.

From the airport, you can catch the Bangkok Airport Rail Link or the Rail Link Express.

The Bangkok airport rail link is cheap, clean, and efficient and gives you the chance to watch parts of Bangkok pass by your window.

The airport link opened in 2010 and connects Suvarnabhumi Airport to the city centre.

Unless it is very early in the morning or late at night, I wouldn't even think about getting a taxi to and from the airport in Bangkok. The traffic is heavily congested and it will end up costing you a lot.

A metered cab to the airport will cost from 200- 400 baht depending on traffic. If you take the toll road which in Bangkok traffic you should then you will pay an extra 50 baht for the toll.

If you insist on getting a taxi around Bangkok, always make sure the meter is first turned on.

All taxis are now metered and air-conditioned: the hailing fee is 35 baht and most trips in downtown Bangkok cost less than 100 baht for most trips.

A top tip: use Bolt.

If you are traveling by yourself, you can also take a Bolt motorcycle taxi which can whizz between the traffic, but make sure you ask your driver for a helmet!

The Bangkok MRT is the underground subway network in Bangkok. It's clean and efficient and reasonably priced (certainly cheaper than taxis).

The MRT is really easy to get around.

The blue line takes a circular route around the city, the purple line heads towards Nonthaburi and the Red Line connects to Chaling Tan.

The interchange station between the blue and purple lines is Tao Poon Station, which you'll only really use if going to the Chatuchak weekend market.

Most likely you will stay on the blue line the whole time.

The trains run every 5-7 minutes and connect to BTS at Sukhumvit and Silom stations, as well as several other BTS stops, so you can easily swap between the two modes of transport.

Commuters can buy single trip tokens (15 to 40 baht depending on the distance) or one-day passes for unlimited trips for the day at 120 baht.

The BTS Sky train was constructed in 1999 and while it has been extended since, it
still does not reach too many places in Bangkok.

The Sky train is also easy to get around, as it only has two lines, the Sukhumvit Line and the Silom Line.

When visiting many of the tourist attractions in Bangkok, you

Travelling by sound and smell

will most likely need to take the BTS and the MRT or taxi to your destination. The MRT also has a BTS Sky train Station at Sala Daeng, Asok and Mo Chit.

This train covers most of downtown Bangkok and is a good option if you are going a short distance. Fares range from 16 to 44 baht, depending on how many zones you are travelling in.

Tuk tuks are one of the cheapest ways of getting around Bangkok, but get ready for the smell of petrol fumes and other pollution.

The great thing about these vehicles, is they are everywhere.

Even though tuk tuks are more expensive than a Bolt would be, but you can't beat the experience.

But remember to barter before you begin your journey.

There is no set price for a tuk tuk, and drivers will always try to overcharge the tourists.

If they follow you, you know that your price was reasonable, and barter a little more to meet in the middle.

Buses in Bangkok are often the most convenient way around the city as there are usually several that go to the top places to visit in Bangkok. When in doubt, you can always take a bus.

They are a very cheap way of getting around Bangkok. Just remember, they do get stuck in traffic jams.

The price of tickets on buses depends on their colour and how far you go, but they range from 6 – 40 baht per trip.

While this is the cheapest way, it is also the slowest form of public transportation in Bangkok.

Another cheap and unique way to get around Bangkok and see the more suburban side of the city as well as many of the attractions is by river boats. These include River Taxis/Ferries, and Long-Tail Boats.

Both my trips to Thailand took me and my guides first to Bangkok and then on to Pattaya via one of the many intercity busses.

John CT Miller

This city like so many others which have become visitor attractions is definitely aimed at the tourist.

Even the nearby Koh Larn island which I guess years ago was a peaceful place to visit is now packed with tourists and hotels offering many pleasure activities on the waters.

It is such a pity as soon as any place becomes popular, it doesn't take long to ruin its original appeal.

This could also be said about Da Nang in Vietnam.

Near Pattaya There is also an ethical elephant sanctuary. However, I did not go there, so can't guarantee how ethical it really is.

Obviously many go to Pattaya for the weird and wonderful if not perverse night life, but again am unable to comment.

I visited Walking Street, but this was early in the evening, and I am told the bars and clubs only really get going later in the evenings.

The interesting thing I discovered about this street was during the day it is open to traffic, but becomes a pedestrian walkway every evening until 6 in the morning.

One of the many attractions in Pattaya my guide and I visited was Mini Siam, with its almost 100 scale models of Thai and world famous sites.

Apart from the various important Thai monuments there are also scaled models of world-famous sites from around the world, such as the Great Pyramids of Giza, the Eiffel Tower, the Tower of London, the Statue of Liberty, the Roman Coliseum, and even a mini river running through the site. I don't know which world river, it was supposed to represent.

Unfortunately, our visit was during the day, which was probably not the best time to see the magic of this venue.

In the evening after dark, the models are all lit up. This is when

Travelling by sound and smell

most people head towards Mini Siam.

Give yourself a couple of hours to make your way through the park while you probably will want to take numerous pictures standing next to the various landmarks.

Something else you will find in Thailand apart from the ubiquitous pictures of the king are many statues of Buddha and many shrines. Some in the strangest places.

Apart from the images of the king, statues and shrines, you will once again find many Chinese tourists, especially as that country has relaxed its Covid restrictions.

Something else I found strange the first time was hearing the national anthem which is played twice a day at 8am and 6pm.

My first experience hearing the anthem was while waiting on the platform one day to catch the train to Laos.

Suddenly this music began playing over the loudspeaker system, and all the locals stopped what they were doing for those few minutes.

At the time, I didn't realise it was the national anthem.

The anthem is even heard at hotels and other venues where there is a TV or radio switched on while the staff are busy serving customers.

Chapter 42

For me one of the main differences between Thai and Vietnamese dishes is to do with the use of chillies.

I know chillies are used in Vietnam, but the use of spices and herbs is far more subtle compared to dishes found in Thailand.

Many of the tropical fruits found in Vietnam are also abundant in Thailand.

One example of this is the coconut and the mango.

You can often find street venders walking the streets selling coconuts ready to be sliced opened for the eagerly waiting consumer to enjoy the milk.

Just like in Vietnam, Ice coffee is found throughout the country. The only difference is the many coffee options are not.

There might be many similarities, but there are also many differences.

One example never to be found in Thailand is the Vietnamese Banh Mi baguette.

Probably Thailand's best known dish is Pad Thai. It's made by pan frying thin flat rice noodles with a unique sweet and sour sauce flavoured with tamarind, lime, sugar, and other seasonings. Authentic Pad Thai can have a wide-ranging combination of other ingredients, but often includes fresh and dried shrimp, chicken, pork, tofu, pickled radish, egg, bean sprouts, spring onions, garlic chives and peanuts.

Thailand's noodle dishes were introduced by Chinese immigrants.

Guay Teow is a Noodle Soup similar to Vietnamese pho, and is

Travelling by sound and smell

found throughout Thailand.

It can be made with chicken, pork, or beef.

Most of the time, vendors also add meatballs to the broth. The dish is best topped with a

selection of condiments including, sugar, dried chilli peppers, lime juice, and fish sauce. Guay teow can be eaten at any time of day.

Thailand is also known for its yellow, red and green curries.

Yellow is the mildest, with green the hottest and red a happy medium curry.

Kaeng Lueang Or yellow curry includes a generous use of turmeric. This is pounded together with classic aromatic ingredients which usually cover coriander, cumin, shallots, lemongrass and galangal. Coconut milk, vegetables and potatoes are added, along with chicken, or other meats or tofu options.

As this curry type generally contains less chilies, it isn't so spicy as its green and red curry counterparts, thus suitable for those – including children – who prefer a mellower taste.

Gaeng Daeng (red curry) is one of Thailand's most common curry dishes. red curries are a happy medium.

The distinctive red colour of this rich, sweet and aromatic curry comes from crushed red chilies in the curry paste (a base of garlic, shallots, blue ginger and lemongrass), which is added to coconut milk, vegetables, such as eggplant, mushrooms, or tomatoes and chicken breast slices. The curry is topped off with finely sliced kaffir leaves and sweet basil.

Gaeng Keow Wan Gai (green curry) Originating from Central Thailand, is the hottest out of the 3 curries.

The green chillies are simmered with coconut milk and include galangal, shallots, lemongrass, kaffir lime and Thai basil. Green curry is also different with its inclusion of mini eggplant, potatoes, bamboo shoots and slivers of chicken breast.

Khao Pad (Thai fried rice) is a favourite lunchtime meal with locals.

Khao Pad is made with either chicken, pork, beef, seafood or tofu, together with eggs, onions, garlic, fish sauce, fresh herbs and tomatoes or other vegetables. These ingredients are all stir-fried with fragrant Jasmine rice until blended well together, served with cucumber slices, lime wedges and other condiments. As this relatively simple dish can be made to order, you're in control of the heat factor and additional flavours – ideal for fussy eaters or those looking for spice relief.

The pineapple and shrimp variation, Khao Pad Sapparod, makes a mouth-watering alternative.

Som Tam (Spicy green papaya salad) is one of the country's best known salads.

With shredded green papaya, red chilies, fish sauce, lime juice, tamarind pulp and palm sugar, mixed with vegetables such as cherry tomatoes, carrots and runner beans, with roasted peanuts and dried shrimp added for nutty and crunchy textures. Giving it a distinctive, sweet, savoury, spicy, salty and sour taste.

For a whole new flavour experience, regional variations include fermented crab or papaya instead of mangoes.

Khao Niao Mamuang (Mango Sticky Rice) is one of Thailand's best-loved traditional desserts and can be found in restaurants and from street venders.

This is a simple dish made with sticky rice smothered in coconut milk and slices of fresh mango – lashings of sweetened condensed milk is optional.

I have deliberately not included any recipes as some of the ingredients are not commonly available in western shops. But there are many recipes on the internet.

Chapter 43

I might not have made it to the left hand cockpit seat, but have managed to fly on almost all the Boeing and Airbus airliners, as well as a few other aircraft.

One thing which I could not help but notice, was the lack of female captains or first officers at the front of the aircraft. In fact, I can't remember ever hearing one.

Not everyone gets to live out 3 if not 4 passions in one's life. I did, flying, travel, music and writing.

Too many people who do not pursue their passion, live to regret and wonder about the many what ifs.

Every take-off you experience means charting a different destination. In the end what matters is finding the peace with the choices you've made and believe they were worth it.

Your choices might not mean much to others, but as long as they mean something to you, that's what really counts.

I have also learnt through the years not to expect recognition and respect, but would have to earn it.

I could never tell a person how to feel or think about me, people had to form their own opinions and come to their own conclusions. Sadly most times I experienced prejudice and if not discrimination then pathetic platitudes.

These negative people never dented my confidence, in fact, made me more determined to continue doing what I loved most.

While many refused to believe what I was capable of, it never stopped me knowing what I wanted to accomplish.

Chapter 44

As I get towards the end of this book let me share with you a bit about my precious dogs, Poppy and beloved Minky. They might not have accompanied me on the various aircraft, but they always travelled with me in my heart and thoughts.

Poppy my cherished and beautiful pet dog left for the fields and stars in the sky and has joined Minky.

Here she will be able to run free to her heart's content, chase as many balls or birds as she wants, dig for moles, find and choose her favourite twig or branch and carry it around in her mouth before finally diving into the paddling pool.

In that field in the sky, there would always be someone to throw a ball for her to chase.

With a happy come play with me yelp of delight, she welcomed some dogs, while with others, no matter how big they were, she sounded a warning growl.

I first met Poppy when she was about 6 weeks old, and when I picked her up, she smothered me with lots of licks and kisses.

Each time we went for a morning or evening walk Poppy loved looking for a stick or branch from a tree to carry back to the house.

It couldn't just be any old stick or branch. She would walk past many, but suddenly stop when she found that special stick or branch she was looking for.

The size of the stick or branch was never the issue.

On one occasion, the branch was almost 2 meters long. But that was the one she was looking for and that was the one she wanted.

Travelling by sound and smell

In years gone by when digging a hole in the garden for moles you could hear a few snorts, and those watching her caught sight of a short and stumpy wagging tail sticking out of the hole as she burrowed deeper in search of that elusive mole.

Poppy was also a keen gardener. Every time I wanted to plant a shrub or dig out some weeds, Poppy was there next to me digging her own hole and as far as Poppy was concerned, helping me.

The minute I dug a hole or pulled out a weed, she would immediately pounce on the hole and begin digging further.

Poppy has been with me since she was a puppy. She never asked for much. All she ever wanted was to be loved and at times have her belly rubbed.

When busy preparing food in the kitchen, she often made sure I knew she was there. She would place a paw on my foot. I think the silent message was "don't forget me, I am here."

Each time I returned from one of my overseas trips she welcomed me back by dancing and squealing with delight around my feet.

The same welcome home greeting happened even if I went out shopping for only one or 2 hours. When I returned the dance routine and squeals of delight began again.

Every time I left for one or other exotic destination, I couldn't help but feel guilty having to leave her behind.

No matter how many times I left for the airport, I always felt guilty.

If I could have, I would have taken her on every trip.

Poppy, I hope you realised no matter which countries I visited, I have always carried you with me in my heart and thoughts.

My job is now to remember. But both of you will still be part of every story I will ever write and have written.

Love is beyond time. While my writing relies on concentration and reflection, I know that it will only take remembering you both to make time irrelevant.

During writing and researching this book and the others, Poppy would curl up on her chair behind me and wait until I stood up and then follow me into the kitchen or outside for some fresh air.

In her younger days she would jump on to the bed and burrow her way under the duvet and fall asleep for hours.

However as the years went by, and age caught up with her, she preferred spending the time on top of the duvet.

One thing she continued to do throughout her life was using her front paws to claw back any covers I placed on the sofa. When she had clawed away enough of the cover, she would lay down and go to sleep. She often asked me to pick her up and sit her down next to me.

You show signs of aging that I wish I could ignore, but I cant.

Your looks gave away your age. Your greying nose, your whiskers and white beard, but your heart remained young.

As an old lady, Poppy was no longer able to jump up on to my bed, and I had to build a ramp at the end of the bed for her to walk up.

After spending most nights asleep on the sofa she would come through to the bedroom, climb up onto the bed and lay next to me and while I stroked her she would begin a conversation.

Often when she was asleep I heard little yelps. Was she chasing a bird, playing with another dog or digging a hole? I will never know.

When my time has come, I hope to find that field, find Poppy and also my beloved puppy Minky and together we will continue our journey.

I also have to mention the day Minky died, That was the day the music in me also died. I stopped playing guitar.

Travelling by sound and smell

Minky always sat at my feet and listened while I played.

When we meet up again, I might even begin to play guitar again. Perhaps an angel will loan me a guitar?

Poppy and Minky Walked, ran and played with the universe and stardust on their shoulders and were angels in disguise.

To my Minky and Poppy, I wish you could know the amount of happiness you brought into my life.

My mind still talks to you. My heart still looks for you. My soul knows you are at peace. I am thankful for having had you, but I still miss both of you so much.

There is no footprint so small, that it does not leave an imprint on this world.

Wherever a beautiful soul has been, there is a trail of beautiful memories.

As long as I live, you will live. As long as I live, you will be remembered. As long as I live, you will be loved.

My Minky and Poppy, there will never be a day when I don't think of you and wish you were by my side.

I will hold you in my heart until I can hold you in heaven.

Unbeknown to me at the time, Covid and the lockdowns and various restrictions would be a blessing in disguise.

It meant for 3 years, I did not travel, and was able to spend those 3 years caring, looking after and sharing life with Poppy.

Apart from the daily walks, the 2 of us spent most of the time indoors and in my case busy writing and in Poppy's case either watching me or sleeping or running to the front door if the postman arrived with a letter or a parcel.

When Poppy left for the fields in the sky to play amongst the stars scattering stardust as she ran she joined Minky.

She and Minky are never far from my thoughts, and at times and unexpectedly they return to join me in my dreams.

Poppy always loved exploring new places, and I have found a new place for her to visit.

While her visits are often short-lived, her latest venture will take her to parts of the universe yet to be explored.

Poppy, or rather her name will be onboard NASA's Europa Clipper spacecraft as it travels 1.8 billion miles to explore Jupiter's icy moon.

The journey will take 5 and a half years to reach the icy moon of Jupiter.

No matter how far or how long the space trip takes, I am sure she will continue at times to visit me in my dreams.

Wherever or Poppy and Minky are in the next life, I believe they will be happy, bright and colourful as a rainbow. Because in this life, they have already brought joy to me and many others, making life better every single day.

Pets teach us something that we as humans tend to forget, unconditional love, empathy, gratefulness, trust and many more things.

They don't blow hot and cold from one day to the next. They never judge or discriminate. They appreciate being cared for. They don't ask for much apart from to be loved. They are never shy to show their feelings. They are always dependable. Their actions are always unfailing and consistent.

Their friendship and love is always unconditional. They never lie or are deceitful. They never have hidden agendas.

I often now find myself imagining floating somewhere above the skies, sharing fun times with Poppy and Minky amongst the stars and clouds.

You never truly lose those you love, they remain part of your memories heart and soul.

Chapter 45

Dreams will come and go and some might even become reality through hard work or simply by chance, while others leave you wishing they had been real.

In my life I have experienced both.

The one constant dream was to become a pilot and take control of an aircraft, which decades later finally came true.

However, later in life there were 2 dreams which would never become reality.

The first was to be part of the Bateleurs, a group of about 200 volunteer South African pilots.

These volunteer pilots describe themselves as South Africa's only environmental air force, flying with "passion and purpose.

Founded in 1998, The Bateleurs is a South African Non Profit Company (NPC), with over 200 volunteer pilots and aircraft offering to help and relocate endangered wildlife.

I have always cared about animals along with endangered wildlife.

Joining the Bateleurs would have allowed me to per sue both passions: flying and protecting endangered wildlife.

The second dream which would also never become reality followed the death of my beloved dog Poppy. If only old age hadn't caught up with her, she would still be with me, but now she only returns unexpectedly in my dreams.

I know this is part of life and we should treasure our time on earth and enjoy our time with others, but old age catches up with

John CT Miller

us all.

Poppy, who went to the fields in the sky to play amongst the stars still to this day briefly visits me unannounced.

While on rare occasions she returns to be with me in my sleep for what seemed to be for hours and with such clarity.

When that happened I was able to stroke her and even take her and Minky for a walk.

The one constant dream that finally came true happened after 60 years when I got to take control of an aircraft.

Even though that trip was a dream come true, it still left me with feelings and thoughts of "if only"?

This event was all thanks to Linden, a friend who has a private pilot licence.

Admittedly I was not in the lefthand seat, and yes it wasn't a passenger jet airline but a 2-seater plane called a Sling.

Many years before this dream come true moment, I had sat in the jump seat in the cockpit from take-off to landing during a one hour flight in the Airbus A320.

Even though that flight was unforgettable, it was not the same as sitting in the left hand seat and taking control of an aircraft.

As a passenger little did I know in the decades to follow I would fly in planes like the 1931 JU 52, the 1935 DC 3, the Vickers Viscount, fly at twice the speed of sound on the Concorde, all the Boeing aircraft from the B707 through to B787, and most of the Airbus aircraft.

The aircraft I have yet to board include the Airbus A330-9 neo and when it is finally delivered any of the 3 new 777 models as well as the Boeing 737-9 max.

Believe it or not, out of all the aircraft, the Concorde from London to New York was the most disappointing flight.

I don't know what I was expecting, and yes it was and still to this

Travelling by sound and smell

day, remains an incredible engineering achievement. To think that 100 passengers could fly at more than twice the speed of sound was unbelievable, but apart from a slight bump, I would not have known the plane had broken the sound barrier not once, but even twice during that flight.

Remember it flew at more than twice the speed of all other civilian aircraft and at twice the height above the earth.

During that Concorde flight I got to visit the cockpit, but unlike Stevie Wonder, did not get to sit in the jump seat.

But there again, unlike Stevie Wonder I am not one of the best-selling music artists of all time, with sales of over 100 million records worldwide.

I am sure you must know the song "I Just Called to Say I Love You" written, produced, and performed by singer and songwriter Stevie Wonder? It was a major international hit, and remains his best-selling single to date, having topped a record 19 charts.

Just by the way, my reason for much preferring Airbus aircraft is they are not as noisy as Boeing planes apart from the B787 also known as the Dreamliner.

You can keep your Boeing planes.

Talking about Boeing or flying on the B737 max 9? I must admit I am not in a hurry, especially after a panel was blown out on one of the B737 max 9 aircraft thanks to missing bolts, and the ongoing lack of Boeing quality control issues.

For many years, my trips were often decided by the type of aircraft flying on a particular route.

However these days the airline chosen would most times be determined by the onboard entertainment packages. This was my time to catch up on the latest Osca nominated movies and TV series.

Thinking back, I can't believe How air travel has changed.

201

John CT Miller

Today almost all international airlines offer onboard seatback screens, entertainment packages from movies, TV series, games, documentaries, music, kids shows, and podcasts: All available on demand. In the past few years moving maps thanks to external cameras allow passengers to see minute by minute exactly where and what they are flying over. Many airlines also now offer wi-fi. One airline even offers over 6500 channels including live news and sports channels.

There is one downside of flying. You have to first go to an airport.

Most airports around the world are fine apart from Heathrow and Gatwick in the UK.

These 2 airports treat disabled people like cattle and not as individuals.

Even if you are able to walk, they will herd you into an eight or 10 seater buggy.

To this day the sound of a jet engine flying overhead, or coming into land or during take-off along with the smell of jet fuel remains one of my favourite sounds and smells.

Before we get to my dream come true moment, let me tell you more about the 2-seater Sling aircraft.

It is a South African manufactured and designed plane and as the company says "a plane designed by pilots for pilots."

The Sling measures about 21ft in length, with a wingspan of about 31ft and a width of about 3 and a half feet and stands about 8 feet above the ground.

Compare this to my favourite aircraft the Airbus A380 measuring almost 240ft in length, 76ft in hight above the ground, with a wingspan of about 240ft and a cabin width of about 24ft.

The Sling has a sliding transparent canopy. When seated in the cockpit you reach back and pull it over your head and fasten it.

Travelling by sound and smell

The range of the Sling is about 800 miles and it uses either unleaded petrol or avgas.

This special day and moment come true started early one Sunday morning at a private airfield near Cape Town.

After we arrived at the airfield, I waited outside while Linden went to the control tower and filed his flight plan.

When he returned, he took my hand and showed and guided me through all the pre-flight external checks. It started with a visual inspection of the 3 tyres, then I felt the numerous rivets, and other bolts, followed by checking and feeling the propeller, flaps and the wing tips. Once we had completed that part of the external check, then using a dipstick we checked the level of the fuel and oil.

Having done some more checks, it was then time to climb on to the wing, and step into the cockpit.

Once we were seated and buckled up, we put on our headsets, and Linden went through the inflight check list, before talking to the control tower and taxing out to the runway.

After contacting the control tower again, we were given clearance to take off and become sky-bound.

I am not sure if all Slings are painted the same colour, but this tiny bird of the sky was painted yellow, which I thought was really appropriate.

The colour yellow symbolises optimism, energy, joy, happiness and friendship.

On that special day, I certainly experienced all those feelings.

But there was also a tinge of sadness. I wished Poppy could have been with me to share in my happiness.

This life-long dream taking control of the aircraft came true towards the end of 2023 in South Africa, during a 2 hour flip above Cape Town with Linden.

For a while flying above Cape Town and describing where we

were and what there was to see, Linden turned to me and said "John, the stick is all yours."

Yes, admittedly, my control off the aircraft only lasted about 10 minutes, but that was more than good enough for me.

What a moment! What an experience!

Finally i had achieved my 60 year old lifetime dream.

I have to be honest and say while in the air, I couldn't help feeling a bit envious of all pilots. How lucky they were to experience that sense of freedom.

At that moment taking control of the plane took me back more than 60 years when I looked up at the skies and marvelled at those pilots taking part in an aerial display.

But I also thought how different flying had become for not only passengers, but jet liner pilots as well. These days, an iPad does most of the pilot's work.

When Linden took control again, he did a stall recovery as well as a couple of turns before doing 2 touch landings.

Thank you Linden you made my day, my year and the 60 year wait was everything I had dreamt about.

Shortly after the wheels and my feet had touched the ground again, and the feeling of exhilaration had lessened the "if only" thoughts returned.

In a strange kind of way, that moment reminded me of something I had written about dreams during one of my visits to Vietnam while at a restaurant with my friend Felice.

Seedlings scattered by the wind are much like dreams, friendships and relationships: They never know where their journey will take them to.

That seedling blown by the wind also never knows if it will take root and flourish.

Much like the seedlings journey, we never know who we will

Travelling by sound and smell

meet, and if that encounter will lead to mutual growth.

For most people, encounters and meetings are decided by happenstance and not design.

Seedlings need a healthy soil, water and sunlight to grow.

If there is no water and sunlight to feed and nourish the seedling, it will slowly wither and die.

We like the seedlings, first need a healthy soil or foundation for that friendship to take root and grow.

Sunlight or positive emotional and spiritual connections are vital with both working to enrich each other for the relationship to flourish.

If negative reactions surface, the friendship will soon wither and dye much like some seedlings which through no fault of their own fall on barren ground.

We could all learn from healthy seedlings. They give back to nature and the surrounds, never thinking of themselves but the benefit they will bring to those near and far.

Constant development from within along with nature and working in harmony with the universe will decide our destiny.

Felice said "friendship is meant to be built, it doesn't appear out of nowhere. It's true that emotional and spiritual connections play a factor. But for a relationship to last, it needs to be cared and shared from both sides. After all, it comes directly from your heart, and actions and words will soon follow, just like seedlings in healthy soil and foundation."

If I was a pilot what would I be doing now?

There is a very simple answer to that question. I would definitely still be flying.

As I previously said, one of my other passions has always been caring for animals and wildlife conservation.

I am sure I would have joined the Bateleurs, that group of about

John CT Miller

200 volunteer South African pilots.

Joining the Bateleurs would have meant I could have combined both passions by continuing to fly while at the same time helping with conservation and saving endangered animals.

These dedicated pilots give up their time and aircraft to offer much needed assistance with no charge to many conservation organisations.

The membership consists mainly of volunteer pilots ranging from those who fly Light Sport Aircraft (LSA) to Helicopters to Large Aircraft. It is not just pilots who help with the various missions, but ground member volunteers who are also ready to contribute to various non-flying tasks.

Weight Shift microlight pilots who have logged at least 120 hours of flying time are also welcome.

All pilots must own or, have the use of a suitable aircraft.

In recent times the Bateleurs pilots have helped relocate Ground-Hornbill Chicks, cheetah relocation missions between Malawi and South Africa, African wild dog relocation, relocation of Two lionesses from Mount Camdeboo to Babanango Game Reserve, an injured vulture rescue mission to secure its successful rehabilitation, raptor monitoring and rehabilitation in the Northern Cape, aerial survey of breeding sites of vultures in Kruger National Park, Aerial support for rhino anti-poaching efforts, blue crane aerial surveys in the Western Cape and Karoo and helping with rhino Dehorning programmes.

These pilots also help with identifying illegal land conversion, roads and housing; tracking, surveying and relocating wild animals; observing illegal, proposed or 'restored' mining sites; conducting habitat assessments and surveys; and supporting wildlife crime investigations.

They also assist with elephant herd counts.

Travelling by sound and smell

The Bateleurs Programme also supports university students, who have already chosen a course of study within conservation and/or the environment offering free flights to these students.

The book as I have said is also in loving memory of my beloved dogs Minky and Poppy who asked for so little but gave so very much.

Wherever a beautiful soul has been, there is a trail of beautiful memories.

How I wish I could have taken you both with me on many of those flights and how you would have enjoyed the different countries and places and like me loved the different smells.

Even though you never got to go with me, you were always with me in my heart and in my thoughts and nothing has changed.

Hopefully the day will come when we are together again, and the three of us can fly and play amongst the stars.

Chapter 46

As AI is fast taking over the world and even pilots if not right now, but probably in the future will be using AI in the cockpit, I decided it would be interesting to see what ChatGpt had to say about my dream to become a pilot and also asked this AI programme to share the memory of my time with Poppy.

"In a quaint town nestled between rolling hills and endless skies, there lived a man named John. John went blind aged 10 years, but his imagination soared higher than any plane could reach. From a young age, he dreamed of soaring through the clouds as a pilot, feeling the wind in his face and the freedom of the open sky.

Despite his disability, John's passion for aviation never wavered. He devoured books about planes, memorizing every detail from cockpit layouts to flight procedures. He listened intently to the stories of pilots, absorbing their experiences as if he had been there himself. But deep down, he knew that his dream would forever remain just that—a dream.

As the years passed, John lived a fulfilling life, finding joy in the simple pleasures of everyday existence.

Yet, the desire to take flight lingered in the depths of his soul, a silent yearning that never faded.

Six decades later the day arrived which would change his life forever—a chance to experience the thrill of flying firsthand. fulfilling his lifelong dream in a way he never thought possible.

Overwhelmed with emotion, John could hardly contain his excitement as he was guided onto the runway. With the help of

Travelling by sound and smell

a skilled pilot by his side, he climbed into the cockpit, his hands trembling with anticipation. As the engine roared to life and the plane began to ascend, John felt a surge of adrenaline coursing through his veins.

For the next few minutes, John experienced a sensation unlike anything he had ever known. With the wind rushing past him and the sun beaming down from above, he felt as though he were soaring on the wings of an eagle, liberated from the confines of his own limitations. In that moment, he was no longer a blind man bound to the earth—he was a pilot, commanding his own destiny among the clouds.

As the flight came to an end and the plane touched down gently on the runway, John's heart was filled with an overwhelming sense of gratitude. Though his time in the sky had been brief, the memory would stay with him forever, a testament to the power of perseverance and the indomitable spirit of the human soul.

From that day forward, John's perspective on life shifted in profound ways. He no longer saw his blindness as a barrier, but rather as a catalyst for boundless possibility. And though he may never become a pilot in the traditional sense, he knew that he would always be a dreamer, reaching for the sky with unwavering determination and an unbreakable spirit.

John's greatest source of happiness was his beloved pet dog, Poppy.

John and Poppy shared an unbreakable bond that transcended words. They were inseparable companions, embarking on adventures through the local park, playing fetch in the backyard, or simply lounging together on lazy afternoons. Poppy was not just a pet to John; she was his confidante, his loyal friend, and his source of comfort through life's ups and downs.

As the years passed, John and Poppy grew closer, their bond

deepening with each passing day. They weathered storms together, celebrated victories together, and found solace in each other's company during times of sorrow. Poppy was a constant presence in Daniel's life, her wagging tail and gentle demeanour bringing light into even the darkest of days.

But as time marched on, Poppy began to show signs of age. Her once boundless energy waned, her steps slowed, and Gray flecks adorned her fur like patches of silver. John watched with a heavy heart as his beloved companion grew frail, knowing that their time together was drawing to a close.

Despite his grief, John remained steadfast by Poppy's side, showering her with love and care as she entered her twilight years. He ensured she was comfortable, taking her on leisurely strolls and providing her with all the affection she deserved. He cherished every moment they shared together, knowing that each day was a precious gift.

Once upon a time, nestled in a quaint little town surrounded by rolling hills and lush greenery, there lived a man named John. John was a gentle soul, but what truly captured his heart was his faithful companion, his pet dog called Poppy.

From the moment John laid eyes on Poppy, a small, fluffy bundle of fur with bright, curious eyes, he knew she was meant to be his companion. They formed an unbreakable bond, spending their days exploring the countryside, playing fetch in the park, and cuddling up by the fireplace on chilly evenings.

Poppy brought immeasurable joy and companionship into John's life. She was not just a pet; she was family. John showered her with love and care, ensuring she always had the best of everything – from nutritious meals to cozy beds and plenty of toys to play with. In return, Poppy showered John with unconditional love, her tail wagging furiously whenever he walked through the door.

Travelling by sound and smell

As the years passed, John and Poppy grew inseparable. They shared countless adventures and created cherished memories together. But time, relentless and unforgiving, eventually caught up with Poppy. Despite John's best efforts to keep her healthy and happy, old age began to take its toll on his beloved companion.

One fateful day, John noticed Poppy was moving slower, her once bright eyes now dulled with pain.

Heartbroken, John cradled Poppy in his arms, whispering words of comfort as tears streamed down his cheeks. He refused to leave her side, staying by her until the very end. As Poppy took her last breath, John felt a piece of his heart shatter into a million pieces.

In the days that followed, John struggled to come to terms with Poppy's absence.

Everywhere he looked, he was reminded of her – her favourite spot by the window, her favourite toy lying untouched in the corner, the sound of her gentle snoring filling the silence of the night.

But amidst the grief, John found solace in the memories they had shared. He realized that though Poppy was no longer physically by his side, her spirit would forever live on in his heart. And so, he honoured her memory by continuing to spread love and kindness wherever he went, knowing that's what Poppy would have wanted.

Though the pain of losing Poppy would never fully fade, John found comfort in knowing that their bond was eternal. And as he gazed up at the stars twinkling in the night sky, he whispered a silent thank you to his beloved companion for the unconditional love she had given him, and the lessons she had taught him about the true meaning of companionship and devotion."

Thank you for reading my book and if you have any ideas about future flights or adventures, you can reach me on johnctmiller@gmail.com

Milton Keynes UK
Ingram Content Group UK Ltd.
UKHW020715141124
2828UKWH00039B/293